The

Wisdom

of

Abortion

The

Wisdom

of

Abortion

Its Power, Purpose and Meaning

WisdomOfAbortion.com

The Wisdom of Abortion:
Its Power, Purpose and Meaning

Typeset in Palatino with LaTeX/TeX on Linux

2 3 4 5

ISBN: 1-4116-5452-8

Contents

Chapter 1

Why Abortion?

Thinking about having an abortion? If so, you're in good company. Each year throughout the world forty-six million women choose abortion; that means every two seconds, three women choose abortion. Even in the U.S., where anti-abortion forces are strong, there are 1,370,000 abortions a year, over two a minute.

Already had an abortion? For every hundred women over forty-five years of age in the U.S., forty-three have experienced an induced or spontaneous abortion.

Pregnant but don't want to be? About half of all pregnancies are unintended. And abortion is one of the ways a woman deals with an unintended pregnancy. For every hundred unintended pregnancies, about fifty end in abortion. Of all pregnancies, fully twenty-five in a hundred end in abortion. So anytime you're with a group of women, you may be with someone who has had an abortion.

Why does a woman choose abortion? For serious reasons. No one, not even anti-abortion critics, say the abortion procedure is fun. And abortions as a rule are not free: early abortions generally cost a few hundred dollars; late abortions can run to several thousand. So a woman doesn't choose abortion for frivolous reasons; rather she chooses it as the solution to some pressing problem.

What problem? If you could ask a woman why she chose abortion, what would she say? She might tell you she had an abortion to save her life.

> * I was vomiting all the time and couldn't keep any food down. My doctor told me I had something he called "hyperemesis gravidarum" and if not treated, would lead to dehydration, malnourishment, and possibly heart failure. She said I needed to abort the pregnancy. Otherwise I might die.

Or she might tell you she had an abortion because rape or incest caused her pregnancy. About one abortion in a hundred occurs because of rape or incest.

> * I went to a party with my older sister and was raped by some of the older boys there. I was 13. Luckily my parents found out from my sister and insisted I do a pregnancy test. If they hadn't, I probably wouldn't have found out I was pregnant until I started to show. My parents arranged the abortion. I was scared to go and scared not to. Mom took me. Thank God she did. I can't imagine what it would have been like being a mother at 14.

* I was 18 and pregnant because my boyfriend date-raped me, although they didn't call it that back then. My parents had always been strict Catholics so I didn't want them to know, but somehow my mother guessed. She asked me what I wanted to do and I said I didn't want a child that had been conceived in rape. She didn't say anything for a few days and then gave me the address of an abortion center in another state. She drove me there the next day.

Or she might tell you she had health problem.

* I found I was pregnant when I was 38 and my daughter was 15. I suppose we really didn't think I could still get pregnant and so we weren't as careful as we should have been. I couldn't picture myself almost forty years old giving birth. My health wasn't that good. I worried that the child might be unhealthy, too. I didn't feel I had the strength to raise another child. It wasn't easy but after a short discussion, we decided on abortion.

* I'm seventy-three now. I married young and had three wonderful children. In my early 40's, my husband and I separated. I met another man. A few months later I was shocked to learn I was pregnant. I couldn't imagine having another baby at that age. Could I even carry it to term? I didn't feel like I could. And if I did, would it be deformed somehow? I'd be over 60 when he/she graduated high school. My oldest was pregnant herself. After a little thought, I knew I couldn't do it.

Or she might say her concern was with the health of the fetus. Health concerns cause about six abortions in a hundred.

> * *It was in the late sixties and the father was an acid-head. At the time, there was a lot of misinformation about chromosome damage and I was afraid the kid might be a freak. Turns out the chromosome damage stuff was pure bull, pushed by the same holier-than-thou types who now claim abortion leads to breast cancer.*
>
> * *When I told my boyfriend I was pregnant he said it wasn't his problem. I had just started college and certainly wasn't ready to be a mother. I had been used to drinking a whole lot and the first thing I thought of when I learned I was pregnant was, "What if the baby is deformed?" The abortion was a bit painful but not much and when it was over I felt really glad.*
>
> *That was over ten years ago. I finished school, got married, and have two daughters. During both pregnancies I felt joyful and grateful, unlike my first pregnancy.*

Or she might say she was too young to have a child. So she chose abortion, or her parents chose it for her. This is the case in about ten in a hundred abortions.

> * *I could have been a parent at 15, but chose not to be. I could have brought a child into this world who wasn't really wanted, but chose not to. I could have skipped college and become a mother when I was just a child myself, but decided not to. I'm now in my forties, have a husband*

and three great kids, and know I made the right decision so many years ago.

* I was 14 and pregnant. My mom worked two jobs to support us (me and my two sisters). She helped me get an abortion. Now, I'm 30 and have two great five-year old sons (twins). My husband and I have a reasonably good life and really love our sons. I think I would have been a terrible mom at 16. I'm glad I had my children later when I was mature and married. We shouldn't forced girls to have and raise children before they are ready, when they are little more than children themselves.

Or she might say she turned to abortion because the father didn't want or wouldn't help support the child.

* I had dropped out of high school and had been waitressing for a few years. Lousy job with no health insurance. I started dating the cook and got pregnant. He had no interest in my having a kid, or in me at all once I told him I was pregnant. I couldn't see myself as a single mom. I didn't want to waitress for the rest of my life.

After it was over I took stock of my self and my life. I eventually passed the high school equivalency exam and did a few years of college but didn't finish. I got a job in an office and eventually married. We're not wealthy but we get by. I still work and our two children (our son is four and our daughter almost one) are in day care.

Or she might say there was some other problem with the relationship. Relationship issues cause about fourteen in a hundred abortions.

I was 15 when we started dating. He was 16 or 17. At first things went great but he gradually got real possessive, wanting to always know where I was and who I was with. He didn't like my friends so I gradually stopped hanging around with them. He abused me verbally and after a while I couldn't imagine that anyone else would want me. Eventually he started physically abusing me, slapping or punching me when no one else was around. He never cared about birth control, it was always up to me. Eventually I got pregnant.

My parents hadn't liked him from the first, which of course had made me like him more. But I had no one else to turn to. I was 16 and there was no way they wanted me to have his kid. I think they were afraid that if I did, he'd always be in my life.

If it wasn't for abortion I'll still be tied to the moron I let get me pregnant and I'd have his kid, too. He already had kids from a previous marriage and generally ignored them except to play ball with them now and then. And he'd beat them, too. I was so stupid for letting him use me as his baby sitter, sex partner and maid for so long. I made more money than him, too. When I found out I was pregnant, I finally woke up and smelled the shit. I knew if I had his kid I'd never get rid of him.

Afterwards I felt free in more ways than one. A few years later I met a wonderful guy and got married. We have one great kid and are looking forward to our second in June.

* A few years ago, I was living with friends and working as a waitress. I found myself pregnant by an 18-year-old pothead. I never intended to sleep with him much less let him get me pregnant. Abortion seemed the obvious choice.

Or she might say that she couldn't afford a baby.

* I was 15 and pregnant. My mom had had me when she was 17. I was the oldest. I had two brothers and a sister. We all lived in a trailer. There was no way she wanted me to have the baby at that point in my life. She took me and held my hand the whole time. When it was over, I felt relieved.

Or that she and her husband already had children and couldn't afford another. About twenty in a hundred abortions occur because of financial problems. For every hundred women who choose abortion, sixty of them already have one or more children that need to be supported.

* I was 30 and had a child who was one year old and another who was two and a half. My husband had left and I was trying hard to find enough money to survive. He'd left me pregnant. I had no family to speak of (my parents were deceased) and simply couldn't afford to give birth. I couldn't afford the time away from work (who would care for my children while I was in the hospital?) and couldn't afford another child.

Worse, I'd had problems with my two previous pregnancies and was told the next pregnancy might be worse.

The doctors never said another might be fatal but I couldn't stop worrying about what would happen to my children if I died.

It wasn't easy but after much soul-searching I decided to get an abortion. Afterwards, I felt I had done the right thing. I still feel that way, 18 years later.

* I was 18 and Peter was 21. We'd been married a year and lived in a trailer that didn't have hot water and sometimes didn't even have heat. He worked full-time and I worked part-time. We brought home, I guess, about $470 a week. We made so little, in fact, that the doctor did the abortion for less than his usual fee. Was it the right thing to do? I have no doubt. Eight years later the situation is much better and I'm expecting. We're really excited and looking for to it.

Or she might say she just didn't feel ready to have a child, for one reason or another. About twenty-five in a hundred abortions occur because the woman wants to wait — wait until she finishes college, becomes more mature emotionally, has a more stable relationship, is more secure financially, etc.

* I was 18 and had been with my boyfriend for over a year when I found I was pregnant. We were both in college and knew we weren't ready to have a child. We stayed together afterwards and eventually married, though we had some rocky times along the way. We both had a lot of growing up to do. About five years ago we realized we were ready, financially and emotionally, for children. We

had two planned children and one "accident." They are now 8 months, two and a half, and four, two boys and a girl.

* A few years ago, I found myself pregnant unexpectedly. Our son was 3 years old, my husband was out of work, and money was very tight. And I was on medication for severe depression. It was not the time to have another child. Abortion was the obvious choice.

Ten years later, we have another marvelous son. I have no regrets. At the time, I had my son, my husband and my self to take care of. We just couldn't afford another child then. I have no doubt we did the right thing.

* I guess I was kind of a drug addict when I found myself pregnant. I did a lot of them, anyway. Alcohol, too. I was 19 or 20 and had enough trouble taking care of myself. Can you imagine what kind of mother I would have been? Getting an abortion was a complete no-brainer.

* Raising a child is an enormous task. I've had two abortions. And I'll keep having them if I need to. I'll have kids when I'm in a stable relationship and we're financially ready. Not before.

Or she might say she doesn't want children.

* I've never wanted children. So I've always been very conscientious about birth control. But accidents happen. I've had two abortions.

To save her own life, to end a pregnancy caused by rape or incest, out of a concern for her own health or the health of the fetus, because she's too young, because the father doesn't want a child or because of some other problem in the relationship, because she, or she and her husband, can't afford a child, because she's not ready physically, emotionally or financially — these are some of the reasons you'd hear if you asked women why they choose abortion. These are some of the many reasons that convince a pregnant woman to end her pregnancy through abortion. But do other people find these reason convincing?

Eighty-nine percent of the U.S. public approve of abortion when the woman's life or health is threatened. Seventy-eight percent agree using an abortion is justified to end a pregnancy caused by rape or incest. Almost the same percentage (seventy-six percent) approves if there's a strong chance the baby will be serious deformed or have major health problems. One thing common to all the reasons in this group is that the pregnant woman bears little or no responsibility for being pregnant or for the problem with the pregnancy. Rape or incest isn't the woman's fault. The pregnancy that threatens the woman's life isn't her fault either, nor is the fetus that has a serious health problem. For all these cases, the approval percentage for abortion is high. The situation changes when the woman had sex willingly and can physically bear a child. Within a percentage point or two, forty-two percent of the U.S. public think abortion is justified if the family can't afford any more children, if the women isn't married and doesn't want to marry the father, if there's some other relationship problem, or if the woman wants an abortion for some other reason.

To summarize the public's attitude, if the pregnancy or health problem isn't the woman's fault then it's OK for her to have an abortion. But if the pregnancy is the result of consensual sex and if the woman and fetus are healthy, then all but forty-two percent of the U.S. public believe she should give birth, even if she's a young teenager, even if she'll be a single mother, even if she and her husband can't afford another child.

Less than fifty percent of the public believe abortion is justified when the sexual activity was voluntary and the woman can bear a healthy child. How is it, then, that even after decades of anti-abortion, "pro-life" propaganda such abortions are still legal in the U.S., a democracy where majority rules? If so many people believe certain abortions should be illegal, then why aren't they?

Chapter 2

Well-Being

If so many people believe certain abortions should be illegal, then why aren't they? Studies show that people who oppose abortion are generally older, less educated, and less prosperous than people who support it. Why would younger, more educated, more prosperous people support abortion rights? We'll discuss three possible reasons. First, younger people are more likely to personally know (or be) someone who has had an unintended pregnancy or abortion, so they have a better understanding of the situation than some older people, who grew up before abortion was legal. Second, educated people have a better understanding of the less dramatic reasons for abortion. Reasons in the first group, the group with the higher percentage of public approval, are dramatic: "if I don't get an abortion I may die!" or "I don't want to give birth to such a deformed baby!" or "I was raped!" The second group doesn't have the same drama: "But I won't be able to finish medical school!" But reasons in the second group are just as valid and important as the others, as we'll see. Third, some people owe their education and prosperity to hard work, good fortune — and abortion. Abortion helped them achieve a better life so they naturally support abortion rights for others. For these reasons, and perhaps for others, younger, more educated, more prosperous people often support abortion rights. And because educated, prosperous people are politically influential, abortion is still legal.

> * (Laura) I was in my early twenties and had a job and my own place, and was taking college courses at night. I told my boyfriend I was pregnant and that was pretty much the last I saw of him. Strangely, I had a friend, an acquaintance actually, who was also pregnant. Her boyfriend said he wanted her to have the baby and that he'd marry her. I decided I wasn't ready to go the marriage route; besides, I no longer had anyone to marry anyway. So I chose to get an abortion.
>
> Some months later my acquaintance gave birth. By then her boyfriend had left her, too. She eventually moved to government subsidized housing and got welfare and food stamps. I don't know what happened to her after that but I completed college, got my degree and got a new job. I'm engaged now to a really great guy and look forward to eventually having children.
>
> There's no doubt in my mind I did the right thing.

We can see all three reasons in Laura's account. First, not only was Laura young but she personally knew, and also was, someone with an unintended pregnancy. Second, she was college educated and understood the less dramatic reasons for getting an abortion. Lastly, Laura's abortion decision helped her complete her education and become prosperous.

Laura's decision also enhanced her long-term health in a way that may not be obvious. Laura lives in the U.S., a country that has no national health care system. Therefore, the quality of her health care depends on her income. Because Laura has a professional-level job she has the health care benefits that go with it. She has access to better health care than her less prosperous friend who is on public assistance or now perhaps working a menial job. And when Laura marries and becomes pregnant, she'll have access to better quality pre-natal care and therefore will be more likely to have a healthy child. Throughout her life, she'll have access to better health care than her less prosperous friend.

Laura's decision will have a positive, life-long impact on her well-being, on her education, prosperity, and health. It's not surprising, then, that Laura would be grateful to all those who've fought for abortion rights and would be an abortion supporter herself. And because of her prosperity and health she's in a better position to support abortion rights — with her time and her money — than her less prosperous friend who is fully occupied working a menial job and caring for a child or by now perhaps two or three.

Of course, it's oversimplifying to say educated, prosperous people have abortions and others don't. Many religious people are educated, prosperous and would never consider an abortion. Many political conservatives are educated, rich and against abortion. And many women get an abortion because they can't afford a child, as we've seen.

But it's not oversimplifying to say choosing not to have an abortion often means a woman must leave school and take a low-paying job; and often means the father, if he stays and helps support his child, must do the same. And it's not oversimplifying to say that abortion has the power not only to solve a pressing personal problem but also to improve a person's long-term well-being. Abortion and contraception can help a woman complete her education, stay healthy, and eventually become more prosperous. Though a woman may choose abortion to solve an immediate problem, the beneficial effects of abortion may extend over her entire life.

> * *Back in 1971 the only place to get a legal abortion was New York. My boyfriend took me and was there the whole time. It hurt (but nowhere near as much as giving birth). We stayed together and eventually got married and now have two grown children. Now and then I think we could have had a third child but then I wonder if we even would have stayed together. I don't think either of us was ready at the time. We wouldn't have been able to get our degrees and would be working God knows where. Our life would have been different.*
>
> *We've talked about it now and then and are both glad we did it. At the time, for us it was the right thing to do.*

A woman's decision to limit her number of children, through abortion or contraception, often means a better life not only for her but for her spouse and children, as well. This author's own life is case in point. My grandparents were immigrants who were born in a simple agricultural society. In such a society, a woman marries as young as thirteen or fourteen. Sometimes, her parents choose her husband for her. She often has her first child soon after marriage and continues having children for the next two decades, sometimes as often as one a year. As a child, I had ten aunts and uncles, not counting two that had passed away. But I'm one of only two children. Because my blue-collar parents limited their children to two, with some sacrifice they were able to give us music lessons, ballet lessons, and college. Were I one of four or six children, my life would have been different; I would not have had some of the opportunities I did.

My grandparents had a total of fourteen children. That number may seem huge today but it averages to seven children for my father's parents and seven for my mother's. Yet even today in some agricultural societies, women have as many as fifteen children or more. They stop having children only when they grow too old or when they die, often in childbirth. In such societies, there is little time or money for higher education. Parents are too busy having and trying to feed their children. Education is a luxury beyond their means.

When a nation evolves from an agricultural society to an industrial or technical society, families migrate to the city. In the city, the person with little education finds it difficult to obtain anything better than low-paying menial work. Feeding and supporting a large family is a burden, more so for the person or couple with low-paying jobs. The parents come to see education and birth control as the road to a better life, for themselves or, at least, someday for their children.

> * *I grew up in an extremely poor neighborhood. Many of the women had been deserted by the boyfriends and husbands who fathered their children. Other men beat their women. I knew only a few men that I'd call halfway decent. It seemed to me that the way out of there was education and a good job. I'd been blessed with above average intelligence and planned to make the most of it. When I found myself pregnant at 17 there's was no question in my mind what I'd do.*

Because my parents were born in the U.S., they had opportunities such as a free public education. My father attended high school for two years but didn't finish. My mother did not finish either. So, college was out of the question for both of them. Yet they wanted to give their children a better life. So they emphasized education. And they limited the number of children they had. As a result, both of their children went to college. Of course, there are always extraordinary individuals who succeed in spite of all odds, who attend medical school at night while caring for three children, five younger siblings, and an ailing mother as well. But for most people, the way to a better life is through education, which means having fewer children and having them later in life, which means, in turn, contraception and, when contraception fails, abortion.

Leaders of state are often educated and prosperous. Therefore, they often understand the power of contraception and abortion. They understand that a state can take the path that

Laura took, or it can take the path of her friend. If a state takes the path of Laura, it promotes birth control so it can increase its peoples' level of well-being, their level of education, prosperity, and health. If a state takes the path of unlimited births it suffers the consequences: a low average level of education and an unskilled, impoverished work force.

China and India are two countries following the path to enhanced well-being, to increased education, prosperity and health. The effects of their progress are being felt world wide, particularly in the U.S. where Chinese goods are in every store and where technical jobs are being outsourced to India. We've all experienced the effects of their progress personally. For example, in the U.S. everyone has gone into a store and left with something labeled "Made in China." As another example, some years ago I called support for my PC and spoke to a man whose accent said "Texan." I called again a year or two later and spoke to a woman whose name and accent said "India." Apparently, the Texan's job had been outsourced to someone in India, someone who was educated enough to do PC support, someone who, in all probability, didn't begin having her babies at fourteen, at a rate of one a year. China and India are making great progress. But how are they doing it? One indispensable means is contraception and abortion. To understand how, let's discuss India in a bit more detail.

Much of India's poverty and sickness are caused by over-population. India has sixteen percent of the earth's population but only two and a half percent of its land. One third of its people are illiterate and live in poverty. Over-population means children grow up in slums, drink filthy water, go hungry, and die of preventable diseases.

From 1991 to 2001, India's population increased by 182 million people; that's half the entire population of the U.S. and more people than in all of Brazil. In 2001, India's population reached one billion. Today, it increases by about 1,815 babies every hour. By 2035 India's population may reach 1.4 billion and surpass even that of China.

India's leaders have long feared that uncontrolled population growth would hinder its evolution into a modern industrialized country. So as early as 1951 they began promoting family planning. At the time, the common belief was that more children meant more people to help support the family. So the government began to tell the other side of the story, explaining the benefits and advantages of smaller families. In the 1960's, it began distributing free condoms. In the 1970's, it even resorted to forced sterilization, a practice it has since abandoned.

I once read an article about birth control in India. The article focused on an isolated village of about ten thousand people, a village where most earn their livelihood through farming or weaving, where many have no electricity in their home. In the village, birth control is largely supplied through the "Butterfly" network, which advertises throughout the province. To join the network, the village's physician had to attend a "crash course" in contraception. Now, he has the network's logo (a butterfly, of course) on posters in his office and on the prescriptions he hands to his patients. On his desk is a jar full of the network's Bull condoms, and another full of its Divine Dancer birth-control pills. Many villagers are eager to take control of their own bodies and improve their lives. But some are still wary. One man, the father of twelve, said he'd never use birth control. And a woman who had only four living children after nine births feared the pills would "collect in your stomach and cause a cancer to grow." Worse, according to university studies, throughout the country only about twenty-five percent of the free condoms are "properly utilized." The other seventy-five percent are used as disposable water containers, as dust covers on gun barrels, by weavers to lubricate their looms, and by contractors who mixed them with concrete and tar to build smoother, crack-resistant roads.

Tradition, ignorance and unfounded fears can be difficult to overcome, so India's birth control program has some problems. But it has some successes, too. India's birth rate is slowing down. Before 1996, its population typically grew by 18 or 19 million a year. In 1996, it grew by only 16 million. And although women continue to marry early, the average age has increased from 17 to 17.8 years, a small but encouraging improvement.

As its birthrate declines, India's prosperity increases. In 1979 the average person in the U.S. earned one hundred and five times what the average person in India earned. In 2005, it had fallen to seventy-five times, again a small but encouraging improvement. By 2050, it's been predicted the number will be only five times.

These improvements are the result of government publicity about the advantages of smaller families, freely available family planning information, contraception and abortion. The improvements are the result of people freed from the burden of supporting yet another child each year, having the time and resources to become educated and improve their lives. They are the result of people understanding the facts and acting accordingly. For example, sterilization is now a popular method of voluntary contraception; in some areas it's been estimated that as many as 40% of couples of childbearing age have chosen sterilization.

Another indicator of a country's progress is its infant mortality rate, which indicates the quality of health and life of infants and the general public, too. But what is the infant mortality rate? It's the answer to a simple question: for every thousand infants born, how many die before their first birthday? More generally, it's a measure of how safe it is to have a baby. The lower the rate, the safer it is to have a baby. The higher the rate, the less likely it is that the infant will live to see her first birthday.

But a low infant mortality rate indicates more than how many babies per thousand reach their first birthday. It also indicates the health of general public, too. Why? Because infant health and survival depend very heavily on the mother's health. So a low rate of mortality indicates well-fed mothers who have adequate access to health care, which in turn indicates a somewhat wealthy country. Conversely, a high mortality rate indicates malnourished women with limited access to prenatal and other health care, women who give birth to underweight, sickly babies who die before their first birthday. This, in turn, indicates a country that is poor and undeveloped.

So another measure of India's progress toward becoming a modern country is its infant mortality rate. In 1980, it had an infant mortality rate of 113 deaths per 1,000 births. In 1983, it set the goal of reducing the rate to less than 60 deaths. In 1990 its rate was 84 deaths per 1,000. In 2000 it was 68. In 2005, the rate was 56.3. Of course, these numbers are averages over all of the country. In some areas, infant mortality is still high. For example, the woman with nine births but only four living children probably lives in an area of high infant mortality. In other areas the rate is approaching the low rate of modern counties. As it progresses on the road to becoming a modern, industrialized country India's infant mortality rate will decrease and eventually equal the rate in modern countries — rather, the rate of modern countries that don't have a politically powerful anti-abortion movement.

In the U.S., which has a powerful anti-abortion movement, having a baby is riskier than having one in Cuba, Italy, Greece, Canada or Slovenia. But it is (slightly) safer than having a baby in Lithuania and Croatia. Specifically, as of 2005 for every thousand births, the U.S. had 6.50 deaths, Cuba had 6.33, Italy had 5.94, and Slovenia had 4.45, according the C.I.A. fact book. And Singapore, statistically the safest place in the world to give birth, had 2.29. These numbers show that having a baby in the U.S. is not as safe as having one in Slovenia and almost three times riskier than having one in Singapore.

Why does the wealthiest country in the world have such high infant mortality? One reason is that the U.S., among all the developed countries, has the highest rate of teen pregnancy. But not only does the U.S. proportionally have more ill-prepared poor teenage mothers than any other developed country, it also has a healthcare system that favors the prosperous. So many teenage mothers don't have adequate health coverage and can't afford quality prenatal medical care. So they give birth to underweight, sickly babies that don't live to see their first birthday.

But why does the U.S., of all the world's developed countries, have the highest teen pregnancy rate? One reason is that teens in other countries have better access to sex education and better access to contraception; in the U.S. they have better access to abstinence-only sex education programs and to other anti-contraception, anti-abortion propaganda.

Having a baby in the U.S. is more dangerous that having one in Slovenia partly because of the political clout of anti-abortion groups. Therefore, you might expect the most dangerous place to have a baby in the U.S. would be where the influence of anti-abortion groups is strongest. And you'd be right. The most dangerous place to have a baby in the U.S. is where the anti-abortion forces are most powerful: Mississippi.

Anti-abortion forces are powerful in Mississippi. For example, a news story in 2005 had the headline "Pro-lifers push Mississippi to brink of becoming first abortion-free state in US." It continued: "Outside the last abortion clinic in Mississippi, Melody Miller feels that triumph is within her grasp. A decade ago there were six other such clinics in the impoverished Southern state but they have all closed. Mrs. Miller is determined that this one will meet the same fate, making Mississippi the first state in the union to be abortion-free." And Mississippi's infant mortality rate is the highest in the U.S.: per thousand births Mississippi averages 10.4 infant deaths. That's worse than Latvia's 9.55, Poland's 8.51, and Estonia's 7.87. Mississippi is statistically the most dangerous place in the U.S. to have a baby. (The safest place is Massachusetts, whose 4.9 rate is close to Canada's 4.75.)

In Mississippi, the poor can't get an abortion unless they happen to live near the state's only abortion clinic, or unless they can scrape together enough money to travel out of state. Many can't. So they have a baby they are ill-prepared to care for. Those who would rather not have another child are forced to have the child anyway.

> *I had been waiting for a transplant for over a year. Then I found out I was pregnant. If a donor became available I'd lose my chance if I was pregnant. Besides, the doctors said I'd never be able to carry a pregnancy to full term, anyway. The closest place to get it done was a day's drive. I was taken in an ambulance.*

Forcing people to quit school to have babies creates a state with a relatively unskilled workforce. In a 2003 U.S. Census survey, eighty-eight percent of the people in Wyoming and Minnesota had graduated high school; the state that came in last, where only seventy-three percent of its population had graduated high school, was Mississippi. Forcing teenagers to leave school to support their babies gives Mississippi the lowest percentage of high school graduates in the U.S. And forcing poor people, some without adequate medical care, to bear children gives Mississippi, of all the states in the U.S., the highest infant mortality rate. Further, forcing poor people to have babies they don't want generates more demand for state-provided welfare and medical services. But Mississippi is ill equipped to provide those welfare and medical services because it is the poorest state in the U.S.: it ranks 50th in per capita income; it ranks 50th in disposable personal income per capita; and it ranks 50th in median income of households.

While India and China have wisely decided to control their birth rate and are becoming modern industrial nations, Mississippi seems intent on going backwards, on keeping its birth rate high, its citizens impoverished, and its position behind all the other states secure. It might well take a lesson from China or India.

In this chapter, we've seen some reasons why younger, educated, prosperous people often support abortion. We've seen how abortion and contraception enhance our quality of life, our well-being, our health, education, and prosperity, both on the personal level and on the national level. We've seen how the U.S. infant mortality rate is higher than it should be, partially due to the strength of U.S. anti-abortion forces. We've seen how anti-abortion policies create an unnecessary financial burden for a state or country. And we've seen the connection between a higher rate of infant mortality and policies that some people call "pro-life."

In the next chapter we'll discuss how abortion and contraception save the lives of childbearing women and the general public as well. And we'll see how anti-abortion policies have led to the death of millions of people.

Chapter 3

Survival

In the last chapter we saw how contraception and abortion enhance our well-being and quality of life; we saw how they can help us become educated, prosperous and healthy. In this chapter we'll see how contraception and abortion help save lives, the lives of living, breathing human beings, as opposed to the types of lives anti-abortion people worry about.

To begin, suppose a woman has recently given birth. If she waits at least two years before giving birth again, she increases her second child's odds of survival by fifty percent. So if she uses birth control to space her children then her children are more likely to live. Birth control helps save children's lives. (Yet the Catholic Church, which calls itself "pro-life," forbids birth control and calls it a sin.)

Birth control helps save women's lives, too. In the world, complications from pregnancy or childbirth kill a woman every minute. For every million pregnancies, about a hundred and ten women die in childbirth. But for every million legal abortions, somewhere between ten to twenty women die from medical complications. (Some people dispute those figures and say there are closer to seventy, not a hundred and ten, pregnancy-related fatalities per million pregnancies. Even if they are right, abortion is still three or four times safer than pregnancy.) Abortion is three to ten times safer than pregnancy, and contraception is much safer than pregnancy. Pregnancy can be so dangerous, in fact, that in simple agricultural societies as many as one in sixteen pregnancies end in the mother's death. In developed countries, on the other hand, pregnancy is much safer; for example, in the U.S. only one pregnancy in 3,700 causes the death of the woman. Fewer women die of pregnancy and childbirth in developed countries because they have access to birth control.

But wait! Is it really birth control that reduces the number of women who die in pregnancy and childbirth? Rather, isn't better health care responsible for the lower death rate in developed countries? Yes, the modern health care system of a developed country is more *immediately* responsible than birth control for the lower death rate of pregnant women. But birth control is *ultimately* responsible. Why? Well, how does a country become developed in the first place? How does it acquire a modern health care system? One essential way is by limiting its births, as we saw in the last chapter in the case of India. China is another case in point. China has a strong birth control policy: women who have more than one child are fined, pressured to end the pregnancy, and were once subject to forced sterilization. I don't approve of forced sterilization; it's a human rights violation. But, over all, China's birth control policy has helped it make great progress. A few decades ago the average person used a bicycle to get to work. Now more and more people drive cars; now more and more people have electricity, TVs, and air conditioning in their homes. But controlling its birth rate means more than merely cars and TVs for China's people; it means better medical care, too. Contraception and abortion are helping China develop into a modern country with modern medical care, so that fewer infants die and those that survive have a better life.

So birth control helps a country develop which, in turn, helps it acquire a modern health care system. Birth control helps build a country's health care system another way, too, by helping a person complete the long educational process necessary to become a physician. I've seen this in my own life. Though my blue-collar parents were able to send their two children to college, neither of us became a physician. But another family of four in the neighborhood had a son who eventually became a pediatrician. His father drove a taxi for a living; his mother, as I recall, was a homemaker and had no outside job. Yet, with his family's help he managed to complete medical school. Did contraception help him become a physician? Probably. Certainly had he been one of seven children rather than one of two, his family wouldn't have been able to help him as much. Or suppose before graduating he had fathered a child; then graduating would have been more difficult. In all probability contraception helped him complete the long years of education necessary to become a physician.

Birth control helps saves the lives of women and their children, directly by helping prevent pregnancy and indirectly by helping a country develop a modern health care system. But mothers and children are not the only people whose lives are saved; contraception and abortion have the power to save the lives of other people, too. How? One way is obvious: when a country has a modern health care system everyone benefits. Another way is that they help us avoid overpopulation, a prime cause of war. For example, overpopulation led to World War II and millions of deaths, deaths that might have been avoided with the wise use of birth control.

World War II. We learn about it in school. Six million Jews died. Twenty-five million Russians. Three hundred thousand U.S. military personnel lost their lives. In total it's been estimated an astounding sixty one million people died in that war. But why did they die? What caused the war? Again we learn in school that Germany invaded Poland in 1939, invaded Denmark, Norway and France in 1940, invaded Greece, Yugoslavia and the USSR in 1941, attacked England. But why? The usual answer goes something like this: because a madman who wanted to dominate the world, who lusted for unlimited power, ruled Germany.

But how did the madman convince his country to follow him? Without a following, Hitler would have been merely another lunatic. Every country has lunatics. But unless they convince someone to follow them, they can't do much damage. Hitler did immense damage. How? Because he had the German nation behind him. But how did Hitler convince Germany to follow him? He convinced the German people they needed more "lebensraum" (living space). And why would they need more "living space" than they already had? Overpopulation.

In his infamous book "Mein Kampf," Adolf Hitler observed that Germany's population in 1925 was increasing by 900,000 people a year.   He discussed four ways Germany might accommodate its new members; how it could feed, clothe and shelter its expanding population. The first way was voluntary population control. Hitler rejected this way as violating nature's "survival of the fittest" law. Better, said he, to leave birth unregulated and allow natural competition to weed out the weak from the strong. Let the strong survive and the weak die. The second way Hitler discussed was a higher rate of food production through improved farming techniques. But this he saw as a mere stopgap measure since the population must in time outgrow the ability of the land to feed it. Eventually there would be famine and, as before, the strong would survive and the weak would die. The third way was to increase Germany's productivity though science and technology so it could trade its goods for the food it needed. Hitler scornfully rejected that path: "The talk about the 'peaceful economic' conquest of the world was possibly the greatest nonsense which has ever been exalted to be a guiding principle of state policy." The fourth way of feeding and providing for Germany's mushrooming population was to increase Germany's territory. He wrote: "We must, therefore, coolly and objectively adopt the standpoint that it can certainly not be the intention of Heaven to give one people fifty times as much land and soil in this world as another. In this case we must not let political boundaries obscure for us the boundaries of eternal justice. If this earth really has room for all to live in, let us be given the soil we need for our livelihood. True, they will not willingly do this.

But then the law of self-preservation goes into effect; and what is refused to amicable methods, *it is up to the fist to take*...For Germany, consequently, the only possibility for carrying out a healthy territorial policy lay in the acquisition of new land in Europe itself.... *to obtain by the German sword sod for the German plow and daily bread for the nation."* (Emphasis added.)

Hitler convinced Germany war was necessary. The rest is history, the history of the Second World War, where sixty one million people died, often painfully — sixty one million genuine, living, breathing human beings, not pea-sized embryos. And how many more million were injured? How many more million suffered? Who can say?

For decades, China and India have wisely managed their population growth. Suppose Germany in the decades before World War II had had the wisdom to do the same. Then the need for "lebensraum" (living space) would not have existed; Hitler would have had one less reason to convince Germany to go to war. Had Germany used contraception and abortion wisely, the death of seven million of its own people might have been avoided. Sixty one million people might not have died. In hindsight, which policy would have been genuinely pro-life, Hitler's survival of the fittest policy or a wise, intelligent use of birth control?

Contraception and abortion have the power to save the lives of people other than mothers and their children. By helping nations avoid overpopulation and war, they save the lives of all types of people, young and old, married and single. Not only does avoiding overpopulation help avoid war but it has other benefits, too. When we avoid overpopulation we avoid all the other evils that come with it: hunger and malnutrition, poverty and environmental degradation.

How serious is our overpopulation problem? Well, in 1925, about the time Hitler was writing "Mein Kampf," the earth was home to a bit less than two billion people. By 1950 the number was about two and a half. By 1960, three billion. In 2000, the number was six billion, even though China and other countries had enacted strict population-control laws. So in seventy-five years, world population tripled.

Suppose we wanted everyone in 2000 to be as well housed as everyone was in 1925. Then starting in 1925 for every city, every town, for each and every home that existed, we'd have seventy-five years to build two more. Imagine New York City in 1925. Now imagine building two more New York cities in seventy-five years. Imagine Tokyo in 1925. Imagine Mexico City. Now imagine building two more. In the short space of seventy-five years.

Today, many people don't have adequate housing. Today in the world about one person in six lives in a slum.

Abortion opponents sometimes deny overpopulation is a real problem. They point out that there's still plenty of empty land on earth. This is not an intelligent answer, as they might prove to themselves by merely living on some of that empty land for a day or two. They would soon see that we need more than land to live. We need food, water, housing, medical care, and many other things. Even with today's population, there's not enough food, water, housing, and medical care to go around.

There's not enough food. Malnutrition-related illness kills about ten million people a year. And while our population is mushrooming we are losing land that can grow food. Land overuse and misuse has destroyed the food growing capacity of a third of the earth's soil. And what motivates overuse? Over-population. Even mild malnourishment can stunt the physical and mental development of a child. Yet in U.S. soup kitchens, one person in five is a child. And in the U.S. about thirteen million children each year skip a meal because they don't have the money to buy it. It's been estimated that half the world's children suffer from poverty, war or disease.

There's not enough water. We need water, for ourselves and for the animals we raise as food. But there's not enough for all of us. Over one billion of the earth's people don't have access to safe drinking water. In the world, polluted drinking water kills over 4,500 children a day. In 2004, polluted water killed about 2.2 million people, ninety percent of them children under five years old. At the current rate of water usage, by 2025 five billion people will find it difficult or impossible to obtain clean drinking water.

Further, to grow food we pump huge amounts of irrigation water from underground pools and lakes called aquifers. Decades of continuous overuse have caused available water in aquifers to drop throughout the world. That drop along with pollutants seeping into aquifers means that countries are finding it more difficult to obtain the water they need to grow food. (That's one reason why some scientists are unsure if the earth can support its present population of six billion in the long term.) Some people predict that access to fresh water may soon become a prime cause of war between countries.

We need housing but there's not enough of that, either. About one billion people live in slums, and another billion live in shacks. In third-world countries, one child in three lives in a home that either has a mud floor or has more than five people per room.

There's not enough medical care for all six billion of us. So each year three million newborn babies die who would have survived if they and their mothers had access to clean delivery facilities and low-cost medicines such as antibiotics and tetanus vaccinations. But the women don't have that access, so their children die. Even in the U.S. about forty-three million people have no health insurance; about eight million of them are children.

Today in the world about one person in six lives in poverty; that's one billion people living in poverty. And fully half the people on Earth subsist on less than two dollars a day.

We need a healthy environment but ours is less than healthy. One small example: in the U.S. air pollution caused the number of people with asthma to double since 1980.

We need food, water, housing, medical care and a healthy environment. We want much more. We want cars to drive to work and the gasoline to power them. We want TVs and computers, and the electricity to run them. So we burn coal and oil to generate electricity. But burning coal and oil releases "greenhouse gases" into the atmosphere, which leads to air pollution and global warming. (On August 10, 2003, London reached 100 degrees Fahrenheit for the first time in recorded history.)

Air pollution and global warming threaten our health and in a worse case, runaway greenhouse scenario threaten the very survival of the human race itself.

Even if world population remained static at six billion, we'd have an enormous job ahead of us. To provide a basic level of material necessities for all earth's inhabitants, to avoid more environmental degradation, to avoid a more severe greenhouse effect that, in a worse case, could end human life on earth — even with our present population these are serious challenges. But by 2050 world population is expected to reach nine billion. And China, with almost one fourth of the earth's population, is trading in its non-polluting bicycles for cars.

But what can one person do about all this? What lessons can we learn? One lesson is that the earth already has all the people it needs. And more. So there's no reason to have a child you don't want or aren't equipped to care for. The earth doesn't need any more people. It has quite enough now and expects three billion more by 2050. So if you don't want to have a child now, then don't have a child now. You'll be better off waiting and the child will be better off, too.

Nonsense, cry the anti-abortion people. How, they ask, can a child be better off by not being born? Isn't any life better than none? How, they ask, can you deny the gift of life to a child?

The "denying the gift of life" argument may seem sensible but it isn't. Why? Because anyone who can have a child but doesn't, can be accused of "denying the gift of life." For example, suppose a woman is fourteen and physically able to bear a child. But suppose she doesn't become pregnant when she's fourteen. Then she doesn't have the child that she might have had. So has she "denied the gift of life" to a child? Or suppose a couple uses contraception for a year. Then that year have they "denied the gift of life" to some child? If so, then a Roman Catholic nun who remains celibate also "denies the gift of life" to some child, each and every year of her life between puberty and menopause.

If a woman avoids birth through abortion, through contraception, through the rhythm method, or through abstinence, then what's the difference? In each case, the result is the same: a child that might have been born, isn't. So if we say that one person has "denied the gift of life" to some child then to be fair we must say they all have. But the people who use the "denied life to some child" argument aren't fair. They apply it only when it's useful to them. They don't approve of abortion so they use it for the woman who gets an abortion. In the past, when they were fighting the legalization of contraception, they used it against contraception. (John Calvin, for example, called contraception "the murder of a future person.") But they'd never apply it to a Catholic nun. And of course they'd never apply it to themselves, even if they use contraception.

The "denied the gift of life" argument — if used fairly — applies to everyone except people who are unable have children and the woman who is a "baby factory." The "baby factory" woman begins having babies as soon as she is able and continues having them, one after another, until she can no longer have children. She produces babies as regularly as a factory produces a product.

So suppose one woman avoids having her first child at seventeen through abstinence. And suppose another woman avoids having her first child at seventeen through abortion or contraception. And suppose both have their first child at twenty-seven. Then what's the difference? If at seventeen one woman "denied the gift of life" to some child, then the other woman did, too. The women used different means to achieve an identical result: a child that might have been born was not born.

So should a woman have her first child when she's seventeen and happens to get pregnant by some guy she's been dating a few months? Or would she be wiser waiting until she's twenty-seven and in a stable, loving marriage? What is better for her child?

Should she have her first child when she's nineteen, working as a waitress, with no health coverage? Or after she's finished college, has a professional career and good health coverage? What is better for her child?

Which child is healthier? The one whose mature, stable mother takes care of herself and conscientiously has pre-natal checkups? Or the one whose indigent, teenage mother on public assistance has little or no access to health care? Which child would you rather be? Which child do you want your children to be?

If you don't intend to be a "baby factory", if you intend to have, perhaps, two or three children in your lifetime, then there's no reason to have a child now if you're not ready. The average woman has over twenty or thirty childbearing years. So it's wise to have your children at the time that's best for you. And best for them. Have your children when you're ready. Have them when you're sufficiently mature. Have them when you're emotionally and financially able to care for them. Have them when you're married. There's just no reason a pregnant sixteen or seventeen year old woman should feel obliged to give birth; she'll be much better able to nurture and care for her children later in life.

So have your children when you're best able to care for them, not before. With the way the world is today there's absolutely no reason to have a child now if you don't want to. If you don't have that child, the child will be better off, you'll be better off, and the world will be better off.

> * *There are already enough people in this world. Why bring another one when you're not prepared to care for her? Isn't it better to wait until you can greet your child with joy and gratefulness?*

Where I work, there's a woman I know only by sight; I don't even know her name. For a long time, I'd see her in the halls or the cafeteria. We might nod and smile as we passed, but took no particular notice of each other. One day in the cafeteria I saw her surrounded by friends, laughing cheerfully — and obviously pregnant. Her radiant face told the world she was looking forward to giving birth. It was obvious her child was wanted, would be well cared for, loved and cherished. That's the way it should be for every child. Every child should be a wanted child. The world has no need of any other kind.

Chapter 4

Science

But is it a baby? Is it a human being? If it is, then there is an important difference between the woman who avoids giving birth through abstinence, and the woman who does it through abortion. What does science say? Does science tell us when the fetus becomes a human being? No. Science gives us facts: how large the fetus is, how much it weighs, when it develops various organs, when it can feel pain. But science leaves to us the judgment of when the fetus becomes a human being. Contrary to what some anti-abortion people claim, science does not say the fetus is a human being and does not say human life begins at conception.

When does human life begin? Good question, but first we must answer the more basic question: what do we mean by "human life"? Human life is what pervades the body of a human being; it's what is in every living cell of our body. Each cell of your body has human life. Scrub your face and you scrub off a few cells that possess human life. Cut yourself and you shed a bit of blood that has human life. A woman's unfertilized egg has human life. A man's sperm have human life. But no one worries about the death of the cells you scrub off your face, even though those cells have human life. And no one treats human life as anything sacred. No one gives a funeral to the egg lost in a woman's monthly period. Or to the man's sperm. Not even religious people. As the universe is designed, cells with human life die daily. But no one is bothered by this because what is sacred is not human life but *human lives*, the lives of individual, living, breathing human beings. So, the important question is not when human life begins but when the life of an individual human being starts, when the fertilized egg evolves into a human being.

Unfortunately, there's no clear-cut point where the growing fertilized egg becomes a human being. Rather the evolution occurs gradually over time. It's the same situation when a girl grows into a woman. We might pick some road mark: when she has her first period, loses her virginity, has her first child, on her twenty-first birthday, etc. But any road mark is arbitrary; there just isn't any special point in time where a second before she was a girl and a second later she's a woman.

In this chapter, we'll discuss some of the things science has to say about fetal development. Scientific facts are interesting in themselves, and they help us see through some religion-based arguments against abortion. An ounce of fact can destroy a ton of fantasy.

(*day minus 1*) The human egg is tiny: it takes about a hundred, end to end, to cover an inch. The sperm is even smaller; five hundred of them end to end span an inch.

(Conception) Conception occurs when the egg and the sperm unite, creating a fertilized human egg. Some anti-abortion people insist "human life begins at conception." It does not. The egg and the sperm have human life before conception. Besides, human life — the particular kind of life that pervades all human cells and particularly the fertilized human egg — is not sacred. Rather, human lives — the lives of living, breathing human beings — are sacred. Anti-abortion people confuse the two. They confuse the lives of actual human beings with the "human life" of a fertilized human egg. We'll call this "human lives/human life confusion."

(day 1, week 1) Scientists call the fertilized human egg a zygote. The zygote is the same size as the unfertilized egg, about one hundredth of an inch, smaller than a single grain of salt. Smaller than the period at the end of this sentence.

"Human lives/human life" confusion leads some religions to insist the zygote is a human being, that seven tiny specks (like "........") might be seven tiny human beings in a row. That something smaller than a grain of salt is an actual human being is an idea so detached from reality, so devoid of common sense, it's easy to wonder how anyone can believe it. The answer is people don't believe it because of common sense or facts. Rather they believe it because it's a religious dogma. They believe it because some religious authority told them to believe it.

Religious dogmas don't have to make sense; they don't have to be logical or reasonable. Followers are trained to accept religious dogmas on faith, not logic or reason. In fact, a dogma is a better test of faith if it isn't logical or sensible. If the dogma is logical and reasonable, if it has facts to support it, then faith will have nothing to do. But the dogma that's illogical, the dogma that's unreasonable, is a hurdle that only the true believer can overcome.

(*day 2+, week 1*) What happens to the zygote after conception depends on where conception occurs. Conception can occur inside the woman's body or, with the help of medical science, outside of it.

Suppose the woman's egg is fertilized in the normal place — inside her body. Then about six to ten days after conception, if all goes well, the zygote attaches to the wall of her uterus. But if something goes wrong, then the zygote gets stuck somewhere else. In this case she has an "extra uterine" (i.e., outside the uterus) pregnancy, also called an "ectopic" pregnancy. About two pregnancies in a hundred are ectopic. And for every hundred ectopic pregnancies, ninety-eight are tubal, where the zygote attaches inside the fallopian tube

Ectopic pregnancies are doomed; they never result in a live birth. In an untreated tubal pregnancy, the fetus can eventually burst the woman's fallopian tube, cause her to hemorrhage and threaten her life. The common treatment is to remove the fetus. But some religions call the direct removal of the fetus an abortion and absolutely forbid it, even to save the mother's life. Other religions allow it under special circumstances.

But suppose the woman's egg is fertilized outside her body, in a Petri dish. Then she is probably having an IVF (in vitro fertilization) procedure. Women who have difficulty getting pregnant sometimes turn to IVF. During an IVF procedure, the eggs retrieved from the woman's body are placed in a special growth compound for about five hours. During that time, the sperm are washed and incubated. Then egg and sperm meet in a Petri dish and are left alone so that nature can take its course. After about eighteen hours if some eggs have become fertilized, they are put in another growth compound. Within the next five days, a few fertilized eggs are re-introduced into the woman's body; others are frozen for later use. If all goes well, one or more eggs attach to her uterus, and she becomes pregnant.

The world's first IVF ("test tube") baby, Louise Brown, was born in England in 1978. The first U.S. test tube baby, Elizabeth Carr, was born in 1981 in Virginia. The first pregnancy from frozen embryos occurred in 1983. In the late 1990's a woman gave birth from embryos that had been frozen for twelve years. She had her initial IVF procedure when she was twenty-seven. Eight zygotes were created. Some were implanted. (She eventually gave birth to two baby boys.) The rest were frozen. When she was thirty-nine she decided to try and have more children. So the remaining zygotes were thawed and re-introduced into her body. She again had two boys.

Sometimes the woman who has all the children she wants donates her frozen zygotes to an infertile couple. The zygotes are then implanted in another woman; this "donor embryo" procedure helps otherwise infertile couples have children.

(day 10+, week 2) About this time if the zygote has attached to the uterus, it begins producing hormones. These hormones, which the mother passes in her urine, are what pregnancy tests detect.

The zygote that has attached to the uterus is called an embryo. Some people say a woman is pregnant from the moment of conception; but others use the word "pregnant" only after implantation, only after the zygote becomes an embryo. Implantation occurs six to ten days after fertilization — except in the case of IVF when it occurs anywhere from a few days to twelve years.

How long can the zygote be frozen and still grow into a healthy normal child? At least twelve years, we now know. Some scientists think there may be no limit. The zygote can survive for twelve years, at least, because it's a primitive type of life, like a seed, spore or bacteria. Bacteria trapped over twenty-five million years in amber or crystal have been brought back to life by scientists. Bacteria, seeds and spores are primitive types of life. So is the zygote. In fact, the frozen zygote exhibits no life processes at all, no food assimilation, no excretion, and no growth. It's not a human being, just as a seed is not a tree.

(day 15+, week 3) The embryo is now about 0.08 inches (2 mm) long, microscopic. The "human life begins at conception" religious dogma says that this embryo is as an actual human being. It says that the microscopic zygote in a Petri dish — or the zygote in a refrigerator for twelve years — is an actual human being, too. Even some religious people have difficulty believing this dogma. But they try anyway.

One way religious people try to make this dogma seem sensible is by arguing that the zygote has everything it needs to make a human being, that it only needs nutrients and time to grow. What they say is true; it's a scientific fact. But saying that the zygote has everything necessary to eventually grow into a human being is one thing. Saying that the zygote, right now, is a human being is quite a different thing. An acorn has everything it needs to grow into an oak tree. All it needs are nutrients and time to grow. But an acorn is not an oak tree. The two are very different. A blueprint has everything it needs to become a house or building — all it needs is materials ("nutrients") and someone to build it ("time to grow"). The zygote and the genuine human being are as different as the acorn and the oak tree, the blueprint and the building.

Another way to treat a fundamentally silly dogma is to weaken it a bit: some theologians say that the zygote is an actual human being only if it's free-floating in a woman's fallopian tube. But they say if it's in a Petri dish or a refrigerator, it's not a human being. This idea makes about as much sense as saying when a woman is in one room she's a human being, but when she's in another room she isn't: you're a human being when you're in the living room but not when you're in the kitchen. Whether floating down a fallopian tube or in a Petri dish, the zygote either is a human being or is not. In fact, it is not.

(day 22+, week 4) Going into the fourth week, the embryo is a fifth of an inch in size and has a rudimentary beating heart. Plants don't have beating hearts but animals do. So at this point, we might say the embryo has evolved up the ladder of life from vegetative life to the first stage of animal life. But it doesn't have any brain waves yet.

By the way, we are describing fetal development in days from conception, in what scientists call "fetal age." When fetal age is used the first week begins at conception, the second week begins on the eighth day after conception, etc. Doctors sometimes measure pregnancy a different way, from the first day of the woman's last menstrual period; they call this "gestational age." Using gestational age makes the embryo two weeks older than it actually is. Add two weeks to fetal age to get gestational age.

(*day 29+, week 5*) Going into the fifth week, the embryo looks like a tadpole. It has gills and a tail. It's about the size of a raisin. About twenty-five in a hundred abortions occur before the beginning of the fifth week. "Abortion stops a beating heart," says a popular anti-abortion bumper sticker. At this stage of development abortion stops the beating heart of something about the size of a raisin that looks like a tadpole, has gills but no brain waves.

(*day 36+, week 6*) Going into the sixth week, the embryo is about one half inch (12 mm) long. Its face looks "reptilian." It now has a primitive spinal cord. Buds have formed which will later grow into arms and legs. There are webbed fingers at the end of the arm buds. About forty in a hundred abortions occur sometime before the beginning of the sixth week.

(*day 43+, week 7*) About the seventh week, primitive brain waves can be detected. The embryo still looks like a tadpole. The face has slits that will develop into a mouth and a nose. The embryo is two-thirds of an inch long (head to rump), the size of a kidney bean, and weights 1/25th of an ounce (1 gram), the weight of a grape. About half of all abortions occur sometime before the beginning of the seventh week. Many anti-abortion people have a near hysterical concern with this two-thirds inch, tadpole-like thing with webbed fingers and toes. They believe it's a human being and are very concerned with protecting its life. But they are not so concerned about the lives of genuine human beings. Capital punishment always ends the life of a genuine human being and sometimes ends the life of an innocent one, too. But this doesn't concern them very much. War means the death of enemy soldiers as well as civilians. But no major religion has ever said, "This war is unjust, your religion forbids you from fighting in it." Not even the Christian churches of Nazi Germany.

> *We have four wonderful children. I had thought about it when the fourth came along. Now I was pregnant again. I love my children dearly but we just couldn't afford another. We were just getting by as it was. Jim needed the truck for his business so we decided to sell the car to pay for it. I didn't know what I'd feel like afterwards, but I look at my children and I know I did the right thing.*
>
> *My boss wears his Christianity on his sleeve. He'd fire me in a minute if he knew. Maybe if he paid more than $9.05 an hour we could have afforded another child.*

Can employers who don't pay a living wage genuinely be pro-life? Can religions that have never declared a war unjust call themselves "pro-life"? No, not with any logic or justice.

(day 50+, week 8) Going into the eighth week, the embryo still has a tail. It's about an inch long and weighs less than 1/10th of an ounce (2 grams), the weight of an almond. The brain at this stage is not much bigger than the head of a pin. Fetal brain waves exist but are merely primitive brain stem impulses that control automatic life functions. The brain's cerebral cortex connections don't form until much later, during week twenty. A week or two after that, an EEG can detect activity in the cerebral cortex. During week twenty-three or twenty-four, the brain's cortex/thalamus connections become estab-lished. Most scientists think the fetus can't sense pain before the cortex/thalamus connections are established. Some scientists disagree and believe the fetus can feel pain as early as the twen-tieth week. Neither opinion is popular among pro-life scien-tists, who would like the fetus to feel pain much earlier, as an added argument against abortion. So, some anti-abortion sci-entists claim the embryo can feel pain as early the seventh or eighth week. But it's open to question whether their opinion is honest scientific belief or religiously motivated, anti-abortion propaganda.

(*day 57+, week 9*) About the ninth week, scientists begin call-ing the embryo a fetus. Going into the ninth week, the fetus has developed nostrils. Many major organs have formed but are not yet functioning. The "reptilian" part of the brain has developed but the brain's higher functions have not. The fetus is now less than an inch and a half long and weights about 1/7th of an ounce (4 grams), the weight of about two almonds. Of all abortions, seventy-five in a hundred occur sometime before the beginning of the ninth week.

(*day 64+, week 10*) The fetus is now about one and a half inches long and weighs 1/4th of an ounce (7 grams), the weight of a coin.

Suppose your pregnancy somehow threatens your life. For instance, suppose you have a tubal pregnancy. Some anti-ab-ortion people believe that it is against God's will to destroy a zygote, embryo or fetus, even to save your life. They believe your life is no more valuable than the life of something the size of a grain of salt, a pea or a coin. You may have become pregnant by your husband, by rape, by incest, by accident. You may have other children who need their mother. But no matter. According to these people, God does not want your doctor to save your life. But if you and the fetus die, that's OK with God according to them.

There are many ways a person can be deprived of their common sense and even their sanity, for example, through physical injury or mental illness. Unbalanced, warped religious beliefs can also rob a person of their common sense (and in extreme cases, their sanity). People who have lost their common sense to religion we'll call "religious hysterics." History shows that religious hysterics with sufficient political power pass laws that enforce their manias. It is not inconceivable that someday doctors in the U.S. will not legally be able to end a pregnancy for any reason — not even to save your life. Even if you're a widow with four young children who need their mother.

(*day 71+, week 11*) In week eleven, the fetus can move. It is now about two inches long and has doubled its weight to half an ounce (14 grams). Small plums range from 11 to 25 grams.

An ancient theory describes three stages of fetal development: the vegetative stage, the animal stage, and the human stage. In a crude way the theory agrees with what science tells us. The most basic life processes, which occur even in plants, are food assimilation, excretion and growth. After the zygote attaches to the uterus, it begins to exhibit these most basic, most primitive life processes; it's at the "vegetative stage" of development. Many weeks later it acquires some life processes typical of animals: movement and awareness of the external environment. And brain waves like those of a new-born baby don't appear until the twenty-eighth week, when we might say it has fully reached the human stage. This ancient theory goes by several names, for example, "delayed ensoulment" and "successive animation." We'll meet it again when we discuss abortion and the Catholic Church.

(day 78+, week 12) In week twelve, the fetus is almost three inches long and weighs 4/5th of an ounce (23 grams), a bit less than the weight of four large marshmallows. The fetus now looks something like a human being.

It has always been a standard Christian belief that to be human requires a human soul. For over eighteen hundred years Christianity taught that the human soul doesn't enter the female fetus until about week twelve. So for over eighteen hundred years Christianity taught that the fetus is not a human being at conception, that early abortions do not involve the killing of a human being, as we'll see.

(day 85+, week 13) At the end of week thirteen, a woman reaches the end of her first trimester; her pregnancy is one-third over. The fetus is now about three and a half inches long and has almost doubled its weight again, to one and a half ounces (43 grams), the weight of a candy bar. Abortions during the first trimester cost (in the U.S.) somewhere between two to five hundred dollars, though medical insurance may cover some of the cost. After the first trimester the cost begins to rise, and can eventually reach several thousand dollars. Not surprisingly, the great majority of abortions, specifically, ninety in a hundred, occur during the first trimester.

(day 92+, week 14) In week fourteen, the fingers and toes of the fetus have lost their webs. The fetus is now four inches long and weighs two and a half ounces (70 grams), the weight of a kiwifruit or a very large plum. Of all abortions, almost ninety-five in a hundred (95%) occur before the beginning of the fourteenth week. Of the women who get an abortion after this point, about four in ten are teenagers. Teenagers tend to get late abortions for a variety of reasons. Some of them don't realize they are pregnant, because their menstruation is still irregular or because they can't believe it could happen to them. And some need to get parental approval for an abortion but put off telling their parents. Some wait while they raise the money to pay for the abortion.

(day 99+, week 15) In week fifteen, the fetus is four and a half inches long and weighs about three and a half ounces (100 grams), less than the weight of a small banana. The fetus can now smile, frown, and may have swallowing and sucking movements. So is it now a human being? Or is it a human being only after "quickening," which occurs a few weeks later? Does the fetus become a human being about week twenty-four when most scientists think it develops the ability to feel pain? Or in week twenty-six, when the higher brain activities begin? Or, lastly, in week twenty-eight when fetal brain waves are similar to the brain waves of a newborn baby? It's not an easy question. But it's a question that each woman should be free to decide for herself. For who has a more intimate connection with the fetus than the mother? Who has more intimate knowledge of the fetus than she? Whose wisdom should we trust, the woman's wisdom or the politician's wisdom? Having or not

having an abortion should always be a private, personal decision. Once, India and China forced women who wanted children not to have children, through involuntary sterilization. When are we in the U.S. going to stop forcing women who don't want children to have children, through restricted access to abortion and contraception?

(*day 106+, week 16*) In week sixteen the fetus is five inches long and weighs five ounces (140 grams), about the weight of a medium size lemon. During or after the sixteenth week, the woman first feels the fetus move. Traditionally, the time of first fetal movement was called "quickening," a term based on the old usage of quick to mean alive (as in "the quick and the dead"). Some women describe quickening as feeling like butterflies, belly gurgles, or gas. Quickening can occur as early as the sixteenth week but more commonly occurs between week eighteen and twenty-four. In the past, a woman often didn't tell anyone she was pregnant before quickening. Over the centuries, many Christians, including some popes, said an abortion before quickening did not kill a human being.

(*day 113+, week 17*) In week seventeen, the fetus acquires fingerprints. It's about five and a half inches long and weighs over six and a half ounces (190 grams), about the weight of a Macintosh apple.

> *   *I once read that a fertilized egg has a 50% chance of not making it to term. Many a woman loses a fertilized egg and never even knows she was pregnant. The same article said that even after the woman knows she's pregnant there's a one in five chance of miscarriage, also called spontaneous abortion. It occurred to me that spontaneous*

abortion is the type of event we call an "act of God." I wonder how all those religious nuts would feel if someone told them that God aborts one in five pregnancies! But it's true.

Half of all zygotes are lost, many without the woman ever knowing her egg was fertilized. Can religious people who insist that the zygote is a genuine human being (because it has forty-six chromosomes, because their religion tells them so, or for any other reason) explain why God, day after day, kills so many "genuine human beings"?

(day 120+, week 18) In week eighteen the fetus is six inches long and weighs about eight and a half ounces (240 grams), about the weight of a large apple. The fetus may be sucking its thumb. Sometime between the fifteenth and eighteenth week, a woman may have an amniocentesis, a procedure where a small amount of amniotic fluid is extracted from her womb. Amniocentesis can detect fetal problems such as chromosome abnormalities. A normal human being has forty-six chromosomes. The egg and sperm have twenty-three chromosomes each, so the normal zygote has forty-six.

Because at conception the zygote has all of its chromosomes, some anti-abortion people claim "science says human life begins at conception." In fact, science says nothing of the sort. Rather, it gives us the facts and leaves to us the decision of what is and is not a genuine human being (just as it factually describes the growth of the average teenage girl but leaves to us the judgment of when she becomes a woman). Besides, if forty-six chromosomes make the zygote a human being, then what about people with chromosome abnormalities like Turner's syndrome (forty-five chromosomes) or Down's syndrome (forty-seven)? Anti-abortion people oppose the abortion of a fetus with Turner's or Down's syndrome, too, because their beliefs are based on religious dogma not on any scientific facts about chromosomes. They merely hype scientific facts that happen to agree with their religious dogma, and ignore facts that do not. Besides, unless you have Turner's or Down's syndrome, every cell in your body has forty-six chromosomes. And every cell in your body is alive. But that doesn't make it a human being. Killing a few skin cells when you scrub your face doesn't involve the killing of a human being. Neither does abortion, if done early enough.

(*day 127+, week 19*) In week nineteen the fetus weighs ten and a half ounces (300 grams). It's about six and a half inches long head to rump and about ten inches long head to toe.

Generally, an earlier abortion is safer, cheaper and easier on the body than a late abortion. Why, then, do some women wait? The largest single reason, as we've seen, is that the woman is a teenager who doesn't know or can't admit she is pregnant, or who is afraid to tell her parents. Sometimes a woman has a late abortion because a serious medical problem develops in her or the fetus. Or she spends the time raising money, for the abortion itself or for travel if abortion is not available where she lives.

(*day 134+, week 20*) In week twenty the fetus is ten and a half inches head to toe and weighs 12.70 ounces (360 grams). Some scientists think the fetus can now feel pain; but others believe it can't feel pain until about the twenty-fourth to twenty-sixth week. As a precaution some abortion providers use fetal anesthesia for abortions over twenty weeks. Some states have laws requiring the use of anesthesia after twenty weeks.

(day 141+, week 21) In week twenty-one the fetus is eleven inches and weighs 15.17 ounces (430 grams). For every hundred abortions, ninety-nine are done before this stage.

(day 148+, week 22) In week twenty-two the fetus is almost eleven and a half inches and weighs 1.10 pounds (501 grams).

(day 155+, week 23) In week twenty-three the fetus is almost twelve inches and weighs 1.32 pounds (600 grams). The fetus is viable (can live outside the mother's body) at the twenty-third week, at the earliest.

Abortions during the twenty-third and twenty-fourth week are rare; for each million women who chose abortion only about five hundred and fifty have an abortion during these two weeks of their pregnancy. And many of these abortions are dictated by a medical problem of the mother or fetus.

(day 162+, week 24) In week twenty-four the fetus is over thirteen and a half inches and weighs 1.46 pounds (660 grams). Abortions after week twenty-four are very, very rare; for each million women who chose abortion about two hundred of them have the abortion after that week. So 99.98% of all abortions occur before the end of week twenty-four. This means that of the 1,370,000 U.S. abortions per year, only 274 occur after week twenty-four; of the world's 46 million abortions per year, only 9,200. And often the reason for these abortions is a severe health problem with the fetus.

*   We had tried so hard and so long to have a child. When we found she would die at birth because her spinal cord hadn't developed properly we were devastated. I didn't*

want to live. I was in my third trimester and couldn't bear waiting weeks to give birth knowing she would die anyway. We decided to end the pregnancy, but such late term abortions aren't legal where we live so we traveled to another state, over a thousand miles away.

I was obviously pregnant. I'll never forget the looks of pure hate in the eyes of the pro-life protesters outside the center. They didn't know anything about me so how dare they judge?

(day 169+, week 25) In week twenty-five, the fetus is fourteen inches and weighs 1.68 pounds (760 grams). Most scientists think the fetus first acquires some ability to feel pain between weeks twenty-four to twenty-six, when the brain's cortex/thalamus connections develop. However, physicians find that a fetus born prematurely during that interval often has "significantly less response to pain" than one carried to full term, indicating perhaps the connections are not yet fully formed even between weeks twenty-four to twenty-six.

(day 176+, week 26) In week twenty-six the fetus is almost fourteen and a half inches and weighs 1.93 pounds (875 grams), the weight of a small cantaloupe. At this point, the higher brain functions become active and the fetus becomes conscious of its surroundings. Almost all scientists agree the fetus can now feel some degree of pain.

(day 183+, week 27) In week twenty-seven the fetus is almost fifteen inches long and weighs 2.22 pounds (1005 grams).

(day 190+, week 28) In week twenty-eight the fetus is over fifteen inches and weighs 2.54 pounds (1153 grams), a bit more than the weight of a medium cantaloupe. Now, its brain waves

are similar to those of a new-born baby. Hardly any abortions occur after week twenty-eight that aren't indicated by a health problem of the mother or fetus.

We've now traced fetal development from conception to beyond when over 99.98% of all abortions occur. We've seen what science says and how anti-abortion believers try to twist some scientific facts to their own advantage. In the next chapter we'll see how they do the same thing with verses from the Bible.

Chapter 5

The Bible

Does the Bible say that abortion is wrong? No. Does the Bible say that the fetus is a human being? No, it does not. But in its many hundreds of pages there are a few verses that lend themselves to misinterpretation. In this chapter, we'll examine those verses. We'll see that the verses do not back up the claims of anti-abortion Christians. We'll also discuss Biblical verses that say the fetus is not a human being.

In general, religious followers don't have to justify or explain their beliefs to anyone other than themselves. If they practice their religion in private (and don't harm anyone) they have a right to be left in peace. They are free to decide what is and is not a sin. Different religions have different ideas of what is and it not a sin. To the Jehovah Witness, a blood transfusion is a sin. To a Jewish person, it's a sin to eat meat and dairy together, a cheeseburger, for example. For the Jew and the Muslim, eating ham, pork or bacon is a sin. For a Hindu, eating beef is a sin. Eating a bacon cheeseburger is a sin for all

three religions. As long as religious people practice their religion in private, what they avoid as sinful is their own concern.
If a Jewish person doesn't want to eat a bacon cheeseburger,
that is no one's business but her own.

But once a religious group enters the political arena with
an agenda, once they demand that something be made legal or
illegal, they must be prepared for criticism and disagreement. If
the Jehovah Witnesses campaigned to make blood transfusions
illegal, if Jews, Hindus or Muslims wanted the cheeseburger
banned throughout the country, they would have to give reasons that convince people who don't share their faith. They'd
have to give reasons that convince the general public. And if
their reason is, "God says so" they must able to tolerate the
answer, "You are wrong. God does not say so." If they can't
accept that answer, if they feel it attacks their religion, then they
should keep their ideas to themselves. The U.S. has a secular
government, not a theocracy. So if a religious person can't tolerate anyone disagreeing with their ideas of God and what God
wants, they should keep their ideas out of the political arena.

For decades some Christian groups have worked to get
abortion outlawed. They've given religious reasons for outlaw-
ing abortion, for example, that the Bible says abortion is wrong.
But realizing they must convince people not of their particular
religion, they have also given reasons they pretend are secular:
for example, they say the fetus is a human being and so abortion
is murder. Like anyone else, Christians have the right to fight
in the political arena for what they want. They have a right to
participate in their nation's public debate about abortion. But
once they do, their reasons, their arguments and their claims

become subject to scrutiny and criticism. In this chapter, we are going to scrutinize and criticize some of the Biblical arguments advanced in favor of making abortion illegal. In the next chapter we'll discuss abortion and the Catholic Church. Some readers may see a few of their deeply-held religious beliefs criticized. If you're a Bible-believing Christian, you may see some things that you've been taught thrown into doubt.

Some Christians argue that "Abortion is murder" and that "The Bible condemns abortion." They make these claims in the political arena. Their claims have become part of the public debate on abortion. But is what they say true?

No. The Bible does not condemn abortion. The Bible does not say abortion is wrong.

Ask anti-abortion, Bible-believing Christians if the Bible is God's Word. They'll probably say with great certainty that it is. Now ask if they believe that abortion is against the will of God. Again, they are likely to say it is, with just as much certainty. Now ask them, where exactly does the Bible condemn abortion? Ask them to show you the verses that say God disapproves of abortion. They may show you the Fifth Commandment: "Thou shall not kill." They may say, "No one denies that abortion kills something. Here the Bible says not to kill. So that's where the Bible says abortion is wrong." That's not a sensible answer. Christianity has always approved of killing animals, for food, even for sport. Christians kill insects to help grow food. Christians kill bacteria when they use anti-bacterial hand or dish soap. Christians kill other microscopic things without even knowing it. In the world God has made, it's impossible not to kill. That's why for two thousand years Christianity has said the Fifth Commandment means, "Thou shall not murder; thou shall not kill another human being." So the Fifth Commandment doesn't apply until the fetus becomes a human being.

Besides, you didn't ask them to show you where the Bible says, "Don't kill." And you didn't ask them to show you where the Bible says, "Don't murder." You asked them to show you where the Bible says "Don't have an abortion." or "Having an abortion is a sin." or "The Lord says: 'Don't have an abortion.' " or something similar. They can't show you those verses. Why? Because the verses aren't in the Bible. There isn't a Christian in the world who can show you any Bible verses that plainly and clearly say abortion is wrong. There just aren't any such verses.

Let's put this fact in perspective by looking at some of the things the Bible plainly and clearly says are wrong.

The Bible plainly and clearly forbids wool/linen clothing. Leviticus 19:19 *You shall not let your livestock breed with another kind. You shall not sow your field with mixed seed. Nor shall a garment of mixed linen and wool come upon you.*

The Bible plainly and clearly condemns tattoos. Leviticus 19: 28 *You shall not make any cuttings in your flesh for the dead, nor tattoo any marks on you: I am the LORD.*

The Bible says don't eat pork, ham or bacon. Deuteronomy 14:8 *The pig, because it has a split hoof but doesn't chew the cud, is unclean to you: of their flesh you shall not eat, and their carcasses.*

The Bible forbids various grooming styles for men. Leviticus 19:27 *You shall not shave around the sides of your head, nor shall you disfigure the edges of your beard.*

The Bible condemns charging interest on loans. Deuteronomy 23:19 *You shall not lend on interest to your brother; interest of money, interest of food, interest of anything that is lent on interest.*

But ask a Christian where God in the Bible says something like "You shall not have an abortion." What can they show you? Nothing. Nada. Not a single verse.

This is an amazing fact. Bible-believing Christians say that abortion is against the will of God. Some even say that God hates abortion. But "God's Word" — which in some editions runs five hundred pages, six hundred pages, seven hundred pages or more — doesn't bother to condemn abortion. It condemns a certain type of clothing. It condemns tattoos. It condemns pork. It condemns a certain style of men's grooming. It condemns taking interest for loans. But it doesn't say a blessed thing about abortion being wrong or being against the will of God. Why not? Could it be that when it comes to abortion some Christians don't really follow God's Word? Could it be that they merely follow their own beliefs but pretend they are following God's Word?

The Bible does not condemn abortion. The Bible does not say abortion is wrong. This naturally disappoints many anti-abortion Christians. So they try to provide Biblical support for their anti-abortion beliefs by taking a different approach. They try to show that the Bible says the fetus is a human being. If the Bible did say the fetus is a human being then that, along with the Fifth Commandment, would mean the Bible indirectly forbids abortion.

So, does the Bible say the fetus is a human being? Before we answer that question, notice the "ifs." *If* the Bible says the fetus is a human being. And *if* we combine that with the Fifth Commandment. *Then* we can suppose that the Bible means to say abortion is wrong. This is all very indirect and "iffy." But Bible-believing Christians can do no better. The Bible just doesn't say that abortion is wrong.

So, does the Bible say the fetus is a human being? Yes and no.

In its many hundreds of pages, you can find verses that seem to suggest the fetus is not a human being. And you can find other verses that seem to suggest the fetus is a human being. So based on certain Bible verses, a weak and indirect argument can be made that the fetus is not a human being. And based on other Bible verses, an even weaker and more indirect argument can be made that the fetus is a human being. But, 99.98% of all abortions occur before the end of week twenty-four. And *no* Bible verses say that before week twenty-four the fetus is a human being.

Let's begin with some verses that indicate the fetus is not a human being, that suggest abortion does *not* involve killing a human being. In the Bible, breath, soul and human life are synonymous. A body without breath may be alive but it lacks a human soul and therefore is not a human being. For example, in Genesis Adam becomes a "living soul" in a two step process. First, Adam's body is created, much as the fetus is created in the womb. Genesis 2:7: *And the LORD God formed man of the dust of the ground...* Adam's body is created but he is not yet a "living soul." The verse continues: *...and breathed into his nostrils the breath of life; and man became a living soul.* So only in step two, after God gives him "the breath of life," does Adam become a living soul.

The same idea, that it's the breath that makes us human, appears in Ezekiel 37:8-10, too. The verses describe human bodies that lack breath: *And when I beheld, lo, the sinews and the flesh came up upon them, and the skin covered them above: but there was no breath in them.* As in Genesis, there is flesh and skin but no breath, therefore no human life. The verses continue: *Thus saith the Lord GOD; Come from the four winds, O breath, and breathe upon these slain, that they may live.* So they'll live as human beings only when they possess the breath of life. The verses end: *So I prophesied as he commanded me, and the breath came into them, and they lived, and stood up upon their feet, an exceeding great army.* These verses imply that a body doesn't become a human being until it's animated with the breath of life. The verses contradict the idea that the zygote is human at conception.

Another Biblical indication that God does not consider the fetus a human being occurs in Exodus 21:22-25. The verses begin: *And if men struggle and strike a woman with child so that she has a miscarriage, yet there is no further injury, he shall be fined as the woman's husband may demand of him, and he shall pay as the judges decide.* The Bible says the woman is "with child" (we'll return to this point soon) but the verses show no concern about the death of the fetus. In Biblical times, of course, there were no hospitals, no Intensive Care Units, to save a fetus born prematurely. The miscarriage would inevitably die. Yet the Bible specifies a mere fine for causing a woman to have a miscarriage. And the verses mention "further injury." Since the fetus has miscarried and died, it can suffer no further injury. So who can suffer further injury? Only the woman. So who must have suffered the original injury? The woman. The verses continue: *But if there is any further injury, then you shall appoint as a penalty life for life, eye for eye, tooth for tooth, hand for hand, foot for foot, burn for burn, wound for wound, bruise for bruise.* Why not a "life for life" for the life of the dead miscarriage? Because it's not a human being. Rather it's a fetus. A dead fetus is not a human being according to the Bible so no "life for life" is necessary.

Today, some Christians want laws that give the fetus the same rights as a living, breathing human being — in direct contradiction to what the Bible teaches.

Let's discuss one more verse. Luke 1:31 describes a prophecy of the conception and birth of Jesus. The verse is: *And, behold, thou shalt conceive in thy womb, and bring forth a son, and shalt call his name Jesus.* Now suppose God wanted the Bible to teach the zygote is a human being. Then the verse might have said: "And, behold, thou shalt conceive *a son* in thy womb, and shalt bring him forth, and shalt call his name Jesus." Or better, it might have said: "And, behold, thou shalt conceive *a human being* in thy womb, and bring forth a son, and shalt call his name Jesus." Either of these versions might show God considers the newly-conceived zygote a human being. But neither of these verses is in the Bible.

The verses we've discussed show — in a weak and indirect manner — that the Bible does not regard the fetus as a human being. Now let's examine the even weaker and even more indirect verses that Christians say "prove" the fetus is a human being.

Jeremiah 1:5 has: *Before I formed you in the womb I knew you.*
I've seen this verse on anti-abortion bumper stickers. But why?
What's it supposed to prove? The verse says, "*Before* I formed
thee." God knew Jeremiah before Jeremiah was formed in his
mother's womb. Why? Because God knows everything, past,
present and future. God knows that Alice Jane Smith will be
born a thousand years from now on a small island off the coast
of France. But does that mean that Alice is a human being now,
at this very moment? Of course not. If anything, the verse from
Jeremiah is an argument for reincarnation, not against abortion. Reincarnation is the idea that we exist before we are incarnated in our human body, and that after death we go on to be
incarnated in some other human body. I don't wish to argue for
or against reincarnation. I merely point out that the Jeremiah
verse makes more sense as an argument for reincarnation than
against abortion.

Another argument against abortion can be based on Bible
verses that refer to a pregnant woman as being "with child."
In the Bible, this phrase occurs about twenty-five times. This
argument is not as silly as the argument based on Jeremiah. It
seems to make some sense. But there are at least two problems
with trying to base an argument against abortion on the "with
child" verses.

The first problem is that the verses don't say the fetus
becomes human as soon as the woman's egg is fertilized. Most
people believe the fetus becomes human some time before birth
but the critical question is, "When"? As we've seen, 99.98% of
all abortions occur before the end of the twenty-fourth week.

If the fetus becomes human anytime after that, then the Bible "with child" verses don't apply to 99.98% of all abortions.

So when does the fetus become human? As we've seen, science gives us facts but doesn't directly answer that question. Besides, if science did answer the question but gave an answer anti-abortion Christians didn't like, they'd say science is wrong, just as they do with evolution. Well, when does Christianity say the fetus becomes human? Christianity has always taught that the difference between animals and human beings is that we have souls. Christianity has always taught that a body without a human soul is an animal, not a human being. Further, for over eighteen centuries Christianity taught that the human soul doesn't enter the body until forty to ninety days after conception. So according to traditional Christian belief, only then — forty to ninety days after conception — does the fetus become a human being. Only at that point is the woman "with child." Before that, she is "with fetus."

So the first problem with the "with child" verses is that they do not say the fetus is a human being from the moment of conception. But there's another problem with basing an anti-abortion argument on those verses: the verses need not be interpreted as medically and scientifically accurate. If the ancient Biblical writers didn't have a word for "fetus," they would naturally use a poetic phrase like "with child" to describe a woman who as actually "with fetus." The Bible must be medically and scientifically accurate for the phrase to have any relevance to abortion. But interpreting the Bible that way ignores the last five hundred years of Christian history.

Almost five hundred years ago, Copernicus, the Polish astronomer, said that the earth revolves around the sun. In 1539, Martin Luther convinced himself Copernicus was wrong. Wrote Luther: *People gave ear to an upstart astrologer who strove to show that the earth revolves, not the heavens or the firmament, the sun and the moon... This fool wishes to reverse the entire science of astronomy; but sacred scripture tells us that Joshua commanded the sun to stand still, and not the earth.* Luther "knows" that the earth doesn't revolve around the sun. He "knows" the sun and the moon revolve around the earth. And how does he "know"? He combines two "facts." Fact one: in the Bible, Joshua 10:13 says that God once stopped the sun, not the earth, for a few hours. "Fact" two: the Bible is scientifically accurate.

Joshua 10:13 is not the only Bible verse that says the earth doesn't revolve around the sun. Here's another: Psalm 93:1 *The world also is established that it cannot be moved.* And another: Psalms 104:5 *The Earth is firmly fixed; it shall not be moved.* These verses are in the Bible so how could Luther have been wrong? Luther was wrong because "fact" two — that the Bible teaches literal, scientific truths — is no fact at all. The lesson of the last five hundred years of Christian history is that the Bible is not medically and scientifically accurate.

The Bible says the bat is a bird; scientifically, the bat is not a bird. The Bible says hares chew their cud; they don't. It says not to eat four-legged insects; scientifically, there is no such thing as a four-legged insect. The Bible says that the earth doesn't move and that the sun revolves around the earth. And the Bible in a few verses describes a woman as "with child." Interpreted as medically and scientifically accurate, the Bible is wrong. The earth is not fixed; it spins on its own axis once a day and it revolves around the sun once a year. And a woman isn't "with child," she's actually "with fetus."

Eventually Christians realized the Bible doesn't teach scientific truths. Eventually, they decided "the Bible teaches how to get to heaven, not how the heavens go." In other words, they decided that the Bible teaches spiritual truths, so it's God's Word. But it does not teach medical or scientific truths, so it shouldn't be taken as medically or scientifically accurate. Christians today who base their argument on the Bible's "with child" verses forget the Bible is not medically and scientifically accurate. They might just as well argue — in an equally invalid way — that the earth does not spin on its axis or go around the sun.

We might discuss more Biblical verses that imply the fetus is not a human being, that abortion is not wrong. And anti-abortion Christians could dispute our verses and cite verses that suggest the fetus is a human being. In truth, no Biblical verses are clear and definite. The simple truth is that the Bible says nothing specific about abortion, one way or the other.

This is a curious situation. Christians ignore many explicit Bible prohibitions. Many of the prohibitions are in the Old Testament, but some are in the New Testament. In fact, one is an explicit command of Jesus: Matthew 23:9, *Call no man your father on earth, for you have one Father, who is in heaven.* Most Christians — rightly or wrongly — ignore this direct command of Jesus. But anti-abortion "Christians" are extremely keen on following — and forcing others to follow — a prohibition that is nowhere in the Bible. Could it be anti-abortion "Christians" aren't genuinely interested in following God's Word? Could it be they are only interested in misusing God's Word to give themselves an imaginary divine approval for their own feelings and prejudices?

Confronted with the fact that the Bible doesn't condemn abortion, many anti-abortion Christians will hysterically insist that they *know* God doesn't approve of abortion. ("Knowing" without any supporting facts is a characteristic trait of the religious hysteric.) They'll hysterically insist that God doesn't want any woman to have an abortion. Why then does God cause so many of them? Sometimes a woman's pregnancy ends prematurely. If human intervention ends the pregnancy, it's technically called an "induced abortion." But many abortions are not induced; they are not caused by human intervention. Such abortions are technically called "spontaneous abortions." It's been estimated that half of all pregnancies end before a woman even knows she's pregnant. What ends these pregnancies? What causes a miscarriage? Sometimes a medical problem (a thyroid problem, diabetes, etc.) is to blame. But for every three miscarriages, two are caused by a genetic abnormality of the fetus. Human intervention doesn't cause thyroid problems or diabetes. And it doesn't cause genetic abnormalities. What does?

A miscarriage is a spontaneous abortion. A spontaneous abortion is an "act of God." In the world God has created, fertilized human eggs die each day. Some die because of genetic abnormalities, not human intervention. Many die for no discernable reason: the woman is healthy, the egg is healthy but for some reason fails to implant in the uterus. In the world God has designed, an unsupported heavy object invariably falls to the ground; but in the same world, human pregnancies are not invariably successful. God could have designed the world so a human pregnancy was always successful, but did not. If God doesn't approve of terminated human pregnancies, then why was a world created where so many occur? If God "hates" abortion so much, then why do so many spontaneous abortions occur? Such simple, obvious questions may drive the anti-abortion "Christian" into a unthinking, blind rage (a common response of the religious hysteric). Ask such reasonable questions and they won't see you as having a different opinion but as evil, as in league with Devil (yet another trait of the religious hysteric).

Anti-abortion "Christians" work to deny women access to abortion. Anti-abortion "Christians" force women to have children they don't want, children they are not prepared to have.

* *We already have three children and can't make it on my husband's pay alone. Would the pro-life people help us support a fourth child? I don't think so. Maybe they should worry more about real children who need their help. I wish they'd mind their own business.*

Some women die in childbirth. And what Bible verses can these "Christians" cite to justify their actions? None. The Bible does not say abortion is wrong. The Bible does not say the fetus is a human being. When an anti-abortion "Christian" says that the Bible condemns abortion and that God disapproves of abortion he is indulging in fantasy. What he says is fantasy-based, not reality-based.

In this chapter we've seen that the Bible, clearly and plainly, condemns clothing of a wool and linen mixture, tattoos, eating pork, a certain style of men's grooming, and taking interest for loans. But the Bible does not condemn abortion or even mention it. Why, then, do some religious people condemn it? We return to this question later.

Chapter 6

The Catholic Church

Some people pretend the Bible condemns abortion. It does not. They would like to believe the Bible says the fetus is a human being. But it just ain't so. Some uninformed Catholics would like to believe the same thing. But the Catholic hierarchy knows better. Or, at least, it should. As Pope Paul II wrote in his encyclical letter *Evangelium Vitae* of 25 March 1995, "The texts of Sacred Scripture never address the question of deliberate abortion and so do not directly and specifically condemn it."

Well, if the Bible doesn't condemn abortion how can the Catholic Church? For some Christians it's dogma that the Bible is the only valid source of religious truth. (That's why if they disapprove of abortion, they like to pretend the Bible does, too.) But this Protestant dogma, which goes by the name "Sola Scriptura" ("Scripture only"), is not a Catholic dogma. Rather, according to the Catholic Church, there is not one but three sources of religious truth: Scripture, "authority" (i.e., the Catholic hierarchy of Pope, cardinals, bishops, and priests), and tradition. Scripture does not condemn abortion, as Pope Paul II admits. The Catholic hierarchy does condemn abortion, nowadays, as the killing of a human being; in the past they did not, as we'll see. Does Christian tradition condemn abortion, too? To answer that question, we'll review the historical relation between abortion and the Roman Catholic Church. We'll see that Catholic pronouncements about abortion today contradicted pronouncements of the past.

The question of whether abortion is murder hinges, for Christianity, on another question: does the fetus have a human soul? Christianity has traditionally taught that the soul is what makes us human. Without a human soul a body is merely an animal. So, if the zygote, embryo or fetus doesn't have a human soul it cannot be a human being. And killing what doesn't have a human soul cannot be the killing of a human being, cannot be murder.

When does the fetus become a human being? At conception? At birth? Sometime in between? Centuries before Christianity people wondered about this question. Aristotle was one of the first to answer it. About three hundred years B.C. Aristotle taught that at conception the fetus doesn't possess a human soul, only a "vegetable" soul. In time, the fetus evolves to the level of animal life and receives an "animal" soul. Later, it finally becomes human when the "human" soul enters it. (The human soul is also called the "rational" or "intellectual" soul.)

The time when the fetus acquires a human soul is called "ensoulment." Aristotle said ensoulment occurs in the female eighty to ninety days after conception and in the male about forty days after conception. (He taught that the female fetus matures slower before birth but that women compensate by maturing faster than men after birth.) Aristotle's answer was widely accepted in the ancient world because of his authority and prestige, and the lack of any evidence to the contrary.

The idea that the human soul enters the fetus some days after conception is called "delayed ensoulment." It's also called "successive animation" because the zygote successively evolves through the levels of vegetable life, animal life, and finally becomes animated with human life. The fetus that is ensouled is also said to be "formed" and "animated" (with human life)." In Latin, it's said to be *fetus animatus*. In contrast, before ensoulment the fetus is said to be "unformed" and "unanimated." The unformed fetus, the fetus with a vegetative soul or animal soul, is called in Latin *fetus inanimatus*.

For over eighteen hundred years, the Church made a distinction between *fetus animatus* and *fetus inanimatus*, as we'll see. For over eighteen hundred years the Roman Catholic Church taught the dogma of delayed ensoulment/successive animation. The dogma was taught by popes, by various saints, and by other Church notables. This is an embarrassing fact to the men of the current Catholic hierarchy, so they try to conceal it. Some of them claim that Christianity and, in particular, the Catholic Church has always seen abortion as murder. They do this by quoting certain early Christian documents and individuals. Since Christianity has *not* always seen abortion as murder their arguments, of necessity, are based on misunderstanding or deception.

For example, to support the bogus claim that Christianity has always regarded abortion as murder early Christian writings such as the Didache and the Letter of Barnabas are commonly quoted. But the Didache doesn't say what anti-abortion people would like it to say. It only has: "you must not murder a child by abortion nor kill them when born..." Great idea. But when does the zygote/embryo/fetus become a child? For over eighteen hundred years Christianity's answer was "forty days after conception, at the earliest." Similarly, the Letter of Barnabas has: "Thou shalt not slay the child by procuring abortion." Good sentiment. I'm against slaying children by abortion, too, but the zygote or embryo isn't a child. And the fetus, according to traditional Christian dogma, doesn't possess a human soul until many days after conception.

Besides, the Didache was known to Christians from about 150 A.D. but was not included in the New Testament when the Roman Catholic Church in 397 A.D. formalized the books of the Bible. Why not? Evidently, the Didache lacks divine inspiration. You won't find the Letter of Barnabas in the Bible, either.

Writings like the Didache and the Letter of Barnabas are often quoted to support the fantasy claim that Christianity has "always" said abortion is murder. Individuals commonly quoted to support the same bogus claim are St. Ambrose, St. John Chrysostom, and Tertullian. Of these, Tertullian is unusual in that he really does say what anti-abortion Christians would like him to say: "Now we allow that life begins with conception because we contend that the soul also begins from conception; life taking its commencement at the same moment and place that the soul does." What is not mentioned is that Tertullian eventually became a heretic and left the Catholic Church, which is why no one calls him "Saint Tertullian." Moreover, anti-abortion Christians rarely mention that the individuals and documents they quote expressed a minority view ultimately rejected by early Christianity. They quote mostly from early Christianity's "B-Team." After all, who cares what the non-Biblical Didache or Letter of Barnabas say? Only some anti-abortion Christian trying to prove something false about Christianity and abortion. Early Christianity's "A-Team," on the other hand, taught the majority view: delayed ensoulment/successive animation. This was the view that prevailed; this was the view taught by "Doctors of the Church."

The Roman Catholic Church reserves the official title "Doctor of the Church" for an elite group of thirty-three men and women. Saints Augustine, Jerome, and Aquinas, all Doctors of the Church, taught that the embryo at first does not have a human soul and therefore is not a human being. They taught that ensoulment occurs many days after conception. Augustine taught delayed ensoulment. He writes that the human soul can not exist in a primitive, animal-like body: "The law does not provide that the act of abortion pertains to homicide, for there cannot yet be said to be a live soul in a body that lacks sensation..." Augustine also said that abortion is not murder if the fetus is not "fully formed" with a human soul. And Jerome writes: "The seed gradually takes shape in the uterus, and [abortion] does not count as killing until the individual elements have acquired their external appearance and their limbs."

From at least the time of Augustine, who lived 354-430 A.D., until the 1800's, delayed ensoulment was official Christian belief. Over those many centuries, delayed ensoulment was the settled teaching of the Roman Catholic Church. The Church taught that human life did *not* begin at conception. It taught that abortion before a certain number of days was *not* murder. It taught that the soul did not enter the fetus until forty days after conception for a male, and eighty to ninety days for a female. And if the gender of the aborted fetus was unknown, abortion before ninety days was not considered the destruction of a human being, it was not considered murder. The traditional Roman Catholic teaching that early abortions were not murder influenced the penitentials, the civil laws of Catholic countries, the Church's own Canon law, and the pronouncements of various popes.

The "penitentials" were a type of official Roman Catholic book written as early as 600 A.D. that listed the penances for various sins. Penitentials said sins that prevented conception (birth control and non-vaginal sex) required severe penance, from years to an entire lifetime. But the penance for abortion was only 120 days. But why did the penitentials specify any penance at all if Christianity didn't consider abortion murder? The penitentials specified a penance because Christianity considered abortion the sin of contraception, *not* the sin of murder. The Church wasn't concerned about the life of the fetus. Rather, it was concerned that a woman might use abortion to hide the sin of fornication, adultery or birth control.

Let's digress briefly to discuss the Catholic Church's view of sin. The Catholic Church recognizes two types of sins: venial and mortal. If a Catholic dies with an unforgiven venial sin, she goes to Purgatory for a time but eventually gets to Heaven.

But if she dies with an unforgiven mortal sin, she goes to the place of eternal torture (Hell) prepared by a God who loves her. Along with abortion the Church says that masturbation, contraception, divorce and missing Mass on Sunday are mortal sins.

In some ways the Church sees abortion as a worse kind of sin than the others. But according to the Church the same Hell awaits people who commit the mortal sin of abortion or the mortal sin of intentionally missing Mass on Sunday. (A few decades ago eating meat on Friday was a mortal sin. So was "French" kissing.)

For forgiveness the Church teaches that there must be repentance, there must be regret over the sin. And there must be the sincere intention not to repeat it. Many a Catholic woman who has had an abortion or two feels somewhat sorry and sincerely intends not to have another. According to Church teaching she qualifies for forgiveness. In contrast, numerous Catholic women (and men) regularly commit mortal sins and have no sincere intention of stopping. They masturbate, use contraception, remain divorced, or remain married to a divorced person. Or they regularly miss Sunday Mass. And they lack a prerequisite for forgiveness, the sincere intention not to repeat. Whose soul then, according to Church teaching, is in more danger? End of digression.

The dogma that abortion was not murder influenced the civil laws of many Roman Catholic countries, where early abortion was not a criminal offense and where late abortion was considered a misdemeanor, not a serious crime.

And the dogma was reflected in the Church's own "canon law." Countries have laws; so does the Church. Canon law is the law of the entire Roman Catholic Church as opposed to the civil law of any particular country. In 1140, John Gratian, a monk, compiled "Decretum Gratiani," the first comprehensive collection of the Church's various canon laws. Until 1917 his book (with some additions) was the fundamental compendium of Church law. Gratian's book made a distinction between the abortion of the *fetus inanimatus* and the *fetus animatus*. So, until 1917 official Roman Catholic canon law also distinguished between the *fetus inanimatus* (the unformed, inanimate fetus) and the *fetus animatus* (the formed, animate, human fetus). Until 1917 official Roman Catholic canon law said abortion was homicide only when the fetus was formed.

Even Popes taught that abortion of the *fetus inanimatus* was not murder. For example, about 1211 Pope Innocent III was asked if a monk who had helped his lover get an abortion was guilt of murder, guilty of killing a human being. The Pope said no, because the fetus in question was not yet "animated" with human life. That Pope wrote that the human soul does not enter the fetus until the time of quickening and that abortion before quickening was not murder.

And St. Thomas Aquinas, a Doctor of the Church, wrote in the 1200's that abortion of a fetus that was not yet "animated" with human life was not murder. He said: "The intellective soul, i.e., true person, is created by God *at the completion* of man's coming into being." Aquinas held that the "infusion" of the fetus with a human soul occurred forty to eighty days after conception.

In October 1311, the Council of Vienne met. Attending were four patriarchs, twenty cardinals, one hundred bishops and archbishops, and Pope Clement V. During the Council, the Catholic Church officially endorsed the view of Thomas Aquinas that a true human being is a physical body united with a human, intellectual, rational soul. The Council declared: "...the only begotten Son of God ... *assumed in time* in the womb of a virgin *the parts of our nature united together*, from which he himself true God became *true man*: namely the human, passible *body* and the *intellectual or rational soul...*"

The quote probably needs a bit of explanation. Theologians who tried to understand the Christian dogma of God becoming "man" (i.e. human) found themselves wondering exactly what makes a human, exactly what constitutes human nature. They decided, following Aristotle, Augustine and Aquinas, that a human being consists of two essential parts: the perishable, "passible" animal body *and* the human/rational/intellectual soul. Without the soul, the body is mere animal. They also decided the human soul doesn't enter the body until some days after conception: that "the only begotten Son of God... *assumed in time* in the womb of a virgin the parts of our nature..." Ergo, only the body that possesses such a soul is "true man," truly human.

Two hundred years later, nothing had changed. Pope Gregory XIII (1572-85) said it was not murder to kill a fetus less than forty days old since the fetus "was not human." But in 1588 this settled teaching of the Roman Catholic Church was unsettled for a very short time by Pope Sixtus V, who wrote in the Papal bull *"Effraenatam"* that abortion at any stage of fetal development deserved excommunication and the death penalty. The aberrant teaching of Sixtus was one of the Church's first teachings issued "motu proprio," i.e., without the advice of the rest of the Church. Had Sixtus consulted the rest of the Church they might have protested that his teaching was entirely foreign to traditional Christian teaching. So foreign, in fact, that a mere three years later the next pope, Pope Gregory XIV, restored the ancient, traditional dogma regarding abortion when he declared that only at quickening (which he defined as about 116 days after conception) was the fetus human.

Yet the heresy of Pope Sixtus V lived on. In the 1600's some church and medical authorities began to support "simultaneous animation," the theory that conception and ensoulment occur at the same time. The theory of simultaneous animation says that the animation of the fetus with a human soul occurs at the very same time as conception.

It's interesting that the Church's traditional delayed ensoulment/successive animation teaching describes a process akin to evolution. The body begins with a primitive vegetative soul, gradually evolves to acquire an animal soul, and finally evolves to the point where God gives it a human soul. Similarly, science has found that life on earth evolved gradually, over time. Simultaneous animation, on the other hand, is like the creation story literally-minded Christians read in the Bible where the earth, poof! is created in a day. With simultaneous animation the egg and sperm meet and, poof! a human being is created. Yet even in the Bible, Adam is created in a two-step process. Adam becomes a human being successively, not simultaneously. First God creates Adam's body; then God breathes a human soul into it. Simultaneous animation is a heretical religious theory. Yet in 1869 its advocates prevailed; they overturned the age-old official teachings of the Roman Catholic Church when Pope Pius IX declared that any abortion ended a human life.

Pope Pius IX did not confine himself to condemning abortion. In his famous "Syllabus of Modern Errors" he condemned several other things as well. Let's discuss some of the other things he condemned. Each quote in the following is from the Syllabus. Keep in mind, each quote expresses something that Pope Pius is *condemning* as wrong, as error.

Pope Pius condemns the idea of separation of Church and State, the idea that "The Church ought to be separated from the State, and the State from the Church." In fact, the pope condemns the idea that a country should allow any religion other than Catholicism to exist. He condemns the idea that "In the present day it is no longer expedient that the Catholic religion should be held as the only religion of the State, to the exclusion of all other forms of worship."

Since the pope doesn't believe other religions should be allowed to exist, you might guess he believes an individual person shouldn't be allowed to practice whatever religion she chooses, either. And you'd be right. The pope condemns freedom of religion, he condemns the idea that "Every man is free to embrace and profess that religion which, guided by the light of reason, he shall consider true." The rationale behind condemning freedom of religion is that other Christian religions just aren't as good as Catholicism. Here's another error, according to Pope Pius: "Protestantism is nothing more than another form of the same true Christian religion, in which form it is given to please God equally as in the Catholic Church." And another: "Man may, in the observance of any religion whatever, find the way of eternal salvation, and arrive at eternal salvation." And yet another: "Good hope at least is to be entertained of the eternal salvation of all those who are not at all in the true Church of Christ."

Finally, even the idea of public schools gets Pope Pius' condemnation: "civil society requires that popular schools open to children of every class of the people... freed from all ecclesiastical authority, control and interference..." And "Catholics may approve of the system of educating youth unconnected with Catholic faith and the power of the Church"

Along with separation of Church and State, along with the idea of freedom of religion, along with the idea that other Christian religions are valid ways to God, along with the idea that public schools should be allowed, Pope Pius condemns abortion. Once they've abolished abortion, will some Catholic "pro-life" groups turn their attention to public schools, freedom of religion, Protestantism, and separation of Church and State?

Pope Pius condemned abortion in 1869 but the Church moves slowly; the ancient distinction between *fetus inanimatus* and *fetus animatus* persisted in canon law until 1917. Today canon law only mentions "the fetus." Delayed ensoulment/successive animation is the traditional, age-old, majority teaching of the Roman Catholic Church. Simultaneous animation has traditionally been an unofficial minority view. Yet now it's the current teaching of the Catholic Church. Nonetheless, simultaneous animation is a Johnny-come-lately theory with miniscule support in Church tradition. Some readers may find this astonishing, even unbelievable. But there's no need to take this writer's word for it; look it up. For instance, in the 1992 World Book Encyclopedia article on abortion you can read the following:

> "During the late Middle Ages, abortion before 'quickening' became a generally accepted practice in Europe. Quickening is the time when the mother first feels the fetus moving inside her. Before the 1800's there were few organized religious or legal obstacles to abortion prior to quickening. In the United States, abortion before quickening was not an offence under common law before the 1800's.
>
> Until the late 1800's the Roman Catholic Church had no established doctrine against abortion before animation. Catholic theologians defined animation as the point at which a fetus became human — between 40 and 80 days after conception. But in 1869, the church proclaimed that abortion is never justifiable."

For most of its history, the Catholic Church has said abortion done early enough does not kill a human being.

It would be understandable if some readers cannot believe the truth about abortion and the Catholic Church. Why? Because for decades they've been fed an incomplete, slanted version of Church history. As an example of this slanted Church history, I have before me a "pastoral letter" written some years ago by a Catholic bishop. There's nothing exceptional about the pastoral letter; rather it's a standard example of presenting Church history in an incomplete, slanted way. In the pastoral letter, the bishop mentions the "Sacred Tradition of Faith" and says the Church "does not invent new truths." But, as we've seen, the idea that the fetus is a human being at conception is a new truth, not in accordance with the "Sacred Tradition of Faith."

The bishop then says the Church's condemnation of abortion has never wavered. Here he refers to the Church's condemnation of abortion *as the sin of contraception*. But he allows the reader to misunderstand that the Church has always condemned abortion as the killing of a human being, which it has not. The bishop's letter confuses the Church's traditional teaching about abortion as a sin, with a non-existent tradition of abortion as murder. This sin/murder confusion occurs many times throughout the pastoral letter. (The letter has frequent human lives/human life confusion as well.)

Then the bishop admits: "...it is true that the Middle Ages of the Christian era saw the development of a dispute over the time within which the life in the womb becomes a duly 'formed' child, (i.e. with both body and soul)" but then adds "the philosophical dispute never caused the Church to waver in her absolute rejection of abortion as an option for a Christian." Thus, the Church's centuries-old dogma that the early fetus is not a human being becomes a mere "philosophical dispute." Also, when the bishop writes "absolute rejection of abortion as an option," he merely refers to the Church teaching that abortion was a sin, but again allows the reader to think the Church rejected abortion as the killing of a human being.

The bishop then mentions some of Christianity's "B-Team" documents and individuals that we've already discussed. He

writes that Saint Augustine "severely condemned any deliberate interference with the life of the unborn child." Here's
another example of the sin/murder confusion that runs throughout the entire pastoral letter. The bishop leaves the reader to
believe that Augustine condemned abortion as the killing of a
human being. In fact, Augustine said early abortion was not
murder and merely condemned abortion as the sin of contraception.

Next, the bishop writes: "John Gratian, a Camaldolese monk, in his 'Decretum Gratiani' or 'Concordance of Discordant
Canons' (1140), a private collection of church laws ... offers his
opinion that abortion is not murder if the soul has not been
infused into the fetus (non-animated)." A private collection?
Offers his opinion? Gratian collected the Church's existing
canon laws into one volume, a volume that was the basis of
Roman Catholic Church canon law from 1140 until 1917. But
the bishop paints Gratian as some obscure monk who in "a private collection ... offers his opinion."

Lastly, we've seen how Pope Gregory XIV entirely reversed
the abortion teaching of Pope Sixtus V. But in the bishop's
account, the heretical teaching of Pope Sixtus V is only "slightly
altered" by Pope Gregory.

So with what justification does the Catholic Church condemn abortion as murder? Not with any Biblical justification:
even the Pope admits that Biblical justification is non-existent.
Is there any justification in Church tradition or history? Here,
"non-existent" would be overstating the case. But "minimal
and "insignificant" would be appropriate. So with what justification does the Catholic hierarch say that abortion ends the

life of a human being? With only their own authority: abortion always ends the life of a human being because the Catholic
Church hierarchy nowadays says it does.

Or do they? In 1869 Pope Pius IX declared that although the
fetus is *"not ensouled, [it] is directed to the forming of man. Therefore, its ejection is anticipated homicide."* The fetus is *not* ensouled?
Ejection is not genuine but merely *anticipated* homicide? And
the 1974 Vatican document "Instruction on respect for human
life in its origin and on the dignity of procreation" admits that
the Catholic Church *"...has not expressly committed itself to an
affirmation of a philosophical nature concerning the presence of the
spiritual soul and personhood."* In plain English, as late as 1974
the Vatican was sidestepping the critical question of when the
fetus receives a human soul and becomes a human person.

We've seen that the Catholic Church accepts three sources of
religious truth: Scripture, tradition and authority. Scripture,
as Pope Paul II admits, does not condemn abortion. Traditionally, Catholics have seen abortion, if done early enough, as not
killing a genuine human being. Before 1869 almost all of the
Church's bishops, cardinals and popes taught that early abortion is not murder. Since 1869, some of them have said that
abortion is not genuine but merely anticipated homicide. Do
any of them actually teach that the zygote from the moment of
conception has more than mere "human life" and is a genuine
human being, too? Perhaps they do. Yet they are uncertain as
to when the fetus acquires a human soul, without which the
fetus cannot be a genuine human being. And they have never
said their current anti-abortions teachings are infallible. Even

though the bulk of Catholic tradition, along with the bulk of all the bishops, cardinals and popes who have ever lived, said early abortion is not murder, the men of the Catholic hierarchy nowadays say abortion always ends a "human life." And they say, as they have always said, that it's a sin for a woman to have an abortion. If you find their current opinions and teachings convincing, then don't have an abortion.

Chapter 7

Abortion and You

In the first chapter we discussed various reasons why a woman might choose abortion. We saw reasons range from the dramatic (to save her own life, to end a pregnancy caused by rape or incest) to the ordinary but equally valid (because she's too young, because she already has all the children she can afford, to complete her education). Next, we discussed the connection between abortion and quality of life for a woman and her country. We saw how limiting the number of her children contributes to a woman's well-being; how it helps her become more healthy, educated and prosperous. We saw that entire countries choose to limit the number of births as a way of improving their citizens' well-being and quality of life. Next, we saw how contraception and abortion have the power to increase survival rates of women and children by helping a woman space her births. We explored the connection between abortion, contraception, and the survival of large groups of people. We saw how overpopulation has historically led to war and death, and suffering for millions. We saw that overpopulation today continues to cause the deaths of millions; we saw it causes environmental degradation that, in a worse case, threatens the very survival of the human race. Then we discussed what science has to say about fetal development and abortion. We learned that many abortions occur when the fetus is the size of a raisin, a grape, or a walnut, when it has a tail and gills, before it can feel pain. Then we saw what the Bible has to say about abortion: nothing. Then we saw what the Catholic Church has to say about abortion: a lot, some of it less than honest and frank. We discussed that Church's long history of teaching that abortion is not the killing of a human being, and how it hides its past.

Which brings us to this chapter. We've seen the reasons for abortion, the benefits to well-being and survival, what science, the Bible and the Catholic Church say about it, but what do you have to say about abortion? How do you feel about it?

You may have started this book believing abortion is wrong, read this far, and still believe it's wrong. You may still have religious reasons against it; you may sincerely feel abortion is against the will of God, that it's a sin. Or you may have a gut feeling against abortion that has nothing to do with religion. In any case, if you feel abortion is not a choice you would ever make, no matter what the circumstances, even if you were young and didn't feel ready for the responsibilities of being a parent, even if you already had children and couldn't afford another, even if you had health problems — if you feel that way, then it would be wise to follow your beliefs and feelings; it would be wise never to choose to have an abortion. And I hope you don't feel that this book is meant to persuade you to choose abortion. No, it is meant to give you information about the positive effects and benefits of abortion and contraception. But if that information is outweighed by your own beliefs and feelings then your choice is obvious: if you feel strongly against abortion, then don't have one. The choice is yours; choose what's best for you, whatever it may be. Having or not having an abortion should always be a private, personal decision, and your decision should never be subject to interference or compulsion, like the forced sterilizations once prevalent in India and China, or like the restricted access to abortion and contraception now prevalent in the U.S. If you decide not to have an abortion, you don't have to justify or explain your decision to anyone other than yourself. Do what you feel is right.

The remainder of this chapter discusses practical information for women considering abortion. The last three chapters discuss issues relevant to anyone, regardless of their beliefs and feelings about abortion.

If you're still reading then you probably believe abortion is sometimes a valid option either for yourself or for others. However, for various reasons you may still have some uneasiness about it; we'll discuss two possible reasons, misinformation and indoctrination. For decades anti-abortion groups have spread half-truths and untruths (though they may sincerely believe what they say) about abortion and contraception. Recently, the U.S. government has joined them by promoting school "abstinence-only" sex education classes that give doubtful or flatly wrong information. Over the years you may have absorbed some abortion myths, either from the media or in a class. Here are some of the facts.

If you get an abortion are you more likely to get breast cancer? No, you are no more likely than a woman who never had an abortion.

Are you more likely to have problems with future pregnancies? No.

Are you more likely to have miscarriages later? No.

If you someday give birth is your child more likely to have low birth weight? No.

Is your child more likely to have birth defects? No.

If you've heard differently you'll just have to decide for yourself what to believe. Or you can research the questions, but if you do, pay attention not only to what is said but also to who is saying it. For example, buses on Philadelphia once carried an advertisement that ran "Women who choose abortion suffer more & deadlier breast cancer." Says who? Christ's Bride Ministries, Inc. was the organization that paid for the ad. Meanwhile, the National Cancer Institute, the American Cancer Society, and the American College of Obstetricians and Gynecologists say there's no link between abortion and breast cancer. Who are you going to believe?

> * *My girlfriends did their best to scare the hell out me, to convince me that the abortion would be torture, that I was killing a baby, and that I'd feel guilty and ashamed all my life. I don't know how my boyfriend got me to go. I had a knot in my stomach and felt on the edge of panic, like I wanted to bolt out the door and never stop running. There were two other women in the waiting area. One was like my mother's age and started talking to me. She told me she had done it before and it was no big deal. I got*

the impression she meant more than once. That was when I first began to calm down. I felt myself take a deep breath and the knot let up. I started to cry and my boyfriend put his arm around me.

I chose the twilight anesthetic so I don't remember too much about the abortion itself except that it didn't hurt very much. In fact, I was surprised when they said it was all over. I guess I was waiting for the torture to start.

When it was over, I felt really, really mad at my girl-friends, like they had lied to me and scared me for nothing. I guess it wasn't their fault, though, because they really believed what they said. They just didn't know what they were talking about.

At first, they didn't want to have anything to do with me and that was fine with me. About a year later one came to me when she was pregnant and I helped her get one, too. We became friends again.

Misinformation is one reason you may feel uneasy considering abortion as an option. Early indoctrination is another. Indoctrination occurs when someone gives you biased info with the sole purpose of getting you to believe some doctrine, to accept some point of view. If someone wanted to indoctrinate you against abortion, probably when you were quite young, they would describe abortion as "killing babies." They would not tell you that abortion kills a microscopic-, pea- or walnut-sized fetus that may have gills and a tail, that can't feel pain. On the other hand, if they wanted to indoctrinate you with an accepting attitude towards abortion they would have described it as "saving mommies' lives." Now that you're an adult and have more of the facts you may understand that you were indoctrinated. You may now have a different opinion of abortion. But you may still have some uneasiness about it and feel it is not for you. Early indoctrination can be hard to shake.

If you've already decided to have an abortion, or if abortion is something you're just considering, you may have some questions. In such a situation each woman has her own concerns and questions. We'll discuss a few common ones.

Does it hurt? That depends on how many weeks pregnant you are and whether you take an anesthetic or not. If you're less than nine weeks pregnant then you may be given a pill that induces abortion. If you're less than thirteen weeks pregnant, your doctor may decide on a suction method that's gentle, takes less than ten minutes and doesn't require anesthetic (although if you want it you can usually get it anyway). If you don't choose anesthetic, you'll probably be able to leave the center sooner than if you do. If you do choose anesthetic, and especially if your pregnancy is advanced, you may have to stay overnight. Many women who don't use anesthetic describe the pain as moderate; a few describe it as severe. In either case, it lasts only a short time and hurts much less than giving birth.

> * My abortion didn't take very long. It was a bit painful, but nothing compared to labor and childbirth. I stayed in bed the next day and took some Tylenol. After that, I felt OK.

If you're between thirteen to nineteen weeks pregnant, then you may be given a general anesthetic during the procedure.

If you're over nineteen weeks pregnant, you may have to stay over, for one or two nights, at a medical facility.

Can the fetus feel pain? As we saw in the science chapter, scientists think the fetus can't feel pain until it is twenty-four to twenty-six weeks old. As a precaution, some states mandate anesthetics for the fetus over some number of weeks, twenty weeks, for example.

Is an abortion medically dangerous? There is very little medical risk with abortion performed by competent medical personnel, especially with early abortion. The most common problem is infection, which is treated with antibiotics. If you experience any problems, consult your physician. By the way, it's possible to get pregnant again in as little as two weeks after an abortion, so don't neglect your contraception.

How many days will I be away from work or school? Many women return to work the day after an early, uncomplicated abortion. Some take the day off.

How will I feel afterwards? After the abortion, you may experience feelings of relief and empowerment, or feelings of depression or guilt. Studies indicate many women feel a sense of relief, a feeling of empowerment and self-esteem for having taken charge of their life. In some studies, a very high percentage of women who choose abortion have no regrets and would choose it again in the same situation.

> * *I felt relieved when it was over, like I was in control of my own life again. I guess my ex-boyfriend felt relieved, too, if he felt anything. He already had two kids from two other women and, as far as I could tell, didn't care about them (the kids or the women) or do anything for them*

at all. That's one reason it took me so long to find out they existed. Another reason was that I was really, really stupid. Afterwards, I decided I'd never get involved with a loser like him again.

Negative feelings are possible too, of course. Some women feel generally depressed; some feel sadness and regret or guilt. Some feel angry or suddenly have the need to cry. Some may experience sexual difficulties or difficulties in their relationship. Hormonal changes may cause or intensify these emotions. There are a few situations that may make a woman more susceptible to negative feelings: if she didn't want the abortion but felt coerced, pressured, out-of-control; if she has strong religious beliefs against abortion; if she thought of the fetus as a human being but had the abortion anyway. If any of these situations apply to you, then maybe you shouldn't have an abortion. Other situations that can lead to negative emotions are if a woman has pre-existing depression or other emotional problems, or if health reasons forced her to end an intended, wanted pregnancy

In any case, if you're undecided before the procedure, try to find a genuine counselor who'll help you determine what's best for you. Ask your own inner wisdom what's best for you. Maybe you should have an abortion. Maybe you shouldn't. Only you can decide what's best. Many women don't decide alone. They find it helpful talking to someone. They discuss the question with their friends, their spouse, their parents, or their physician. Some women's clinics have trained abortion counselors you can speak with. There are also many good sources of information, both on the web and in books. Here's two:

Planned Parenthood, 800-230-PLAN,
   http://PlannedParenthood.org/
National Abortion Federation, 800-772-9100,
   http://www.prochoice.org/
What's important is to make the decision that's right for you.

If you do decide to have an abortion, the next thing to do is to find an abortion provider. Many regions in the U.S. don't have an abortion provider, but some of these regions are sparsely populated. Chances are good you'll be able to find an abortion provider close to where you live. But finding a genuine abortion provider is not as easy as it sounds. Some women look in the newspaper or their phone book under "Abortion," "Birth Control," or "Family Planning" and end up at a facility whose only purpose is to prevent them from getting an abortion and to harass them if they do.

One way anti-abortion hysterics use to discourage abortion is by setting up "pregnancy counseling centers" or "crisis pregnancy centers" that dispense slanted information, lies, and intimidation. Such centers try to appear like genuine counseling centers. To attract clients their names often begin with "A" so they are listed first in the phone book. But they usually have only one purpose: to discourage you from getting an abortion. The centers offer free pregnancy tests but may give inaccurate results. If you're not pregnant they may allow you to believe for a while you are, so that you're more open to the "information" they provide, which is actually anti-abortion propaganda. And if you are pregnant they may not tell you, because the longer you wait the harder and more expensive it is to get an abortion. Sometimes none of the staff in a phony counseling center has any medical training. One center in California didn't tell a woman she had a tubal pregnancy. Maybe they genuinely didn't know. When she began hemorrhaging, she had to be rushed to an emergency room.

Some phony centers will have you wait a few hours "until the test results come back." (Test results are ready in a few minutes.) While you are waiting, they'll subject you to anti-abortion propaganda, such as misleading or false claims about the medical dangers of abortion, explicit pictures or films of mutilated fetuses and miscarriages, and stories of women who have had an abortion and say it led to medical and psychological problems, depression, suicidal thoughts, etc. These centers sometimes offer to help you with your pregnancy, emotionally, medically and even financially. They also offer to help you if you choose birth and adoption. Don't believe everything they tell you. Their concern is not you but the fetus you are carrying. After your pregnancy has progressed past the point where you can get an abortion, you may get the promised support but don't count on it. If they were honest, they wouldn't be running such deceptive centers. Some centers will even tell your partner, parents, or employer that you are considering abortion. If you do have an abortion, they may call you on the day you would have given birth and call you a baby killer. One thing you'll never get at these centers is a referral to a place where you can actually get an abortion.

If you want to find a genuine counseling center, a center that will help you decide what's best for *you*, here are a few things you can do. Talk to someone you trust, perhaps someone who has already had an abortion. Or get the name of a local counseling center from Planned Parenthood or The National Abortion Federation, or some other reliable source. Or call a potential center ahead of time and ask them about their services. Ask them if they provide referrals for women who decide to get an abortion. If they hesitate or avoid the question, be suspicious. If they are vague about what services they offer and say "Let's discuss that when you come in" then be suspicious.

Once you have the address of a genuine counseling center, be sure you go to the right place, especially if you're going to an unfamiliar part of town. Phony centers sometime open shop close to genuine centers, hoping to catch the unwary. In Massachusetts, for instance, near a Planned Parenthood counseling center a phony center rented space and put "P.P., Inc." on its door.

When you arrive at a genuine center, you'll probably be given the opportunity to speak with a counselor. If you're sure of your decision before you go to the center, you may be able to skip meeting with a counselor. If you still feel undecided, your counselor will discuss your options and help you make the decision that's right for you. Ask questions and understand your choices. At the center you'll probably talk to a nurse or doctor who will ask you about your medical history. You may be given a blood test and an ultrasound test. The ultrasound test helps determine how many weeks pregnant you are and if the pregnancy is progressing normally. If you decide on abortion, your various options will be explained to you. There are various methods of abortion; which one is used depends on your preference and how many weeks pregnant you are.

Depending on the size of the center, someone may be able to be with you during the procedure.

If you do have an abortion and feel badly afterwards, remember that such feelings often pass in time. If fact, serious psychological problems after abortion are less common than after childbirth. But if you want help, post-abortion counseling is usually available. Many abortion centers can refer you to post-abortion counseling resources. And there are books that discuss emotional healing after abortion. When looking for post-abortion counseling, be careful, just as you were before the abortion. If a center offers counseling for PAS or "Post-Abortion Syndrome," be suspicious; PAS was invented by the same anti-abortion groups that set up phony "pregnancy counseling centers." PAS supposedly describes the effects of abortion on a woman: depression, guilt, uneasiness around children and pregnant woman, sexual dysfunction, drug and alcohol abuse. Over fifteen years ago the American Psychological Association formed a group of experts to investigate if PAS was fact or fantasy. They decided it was fantasy; PAS as a general effect of abortion does not exist. This should be obvious. In the three decades from 1973 to 2002 there were forty-two million abortions in the U.S. So if something like PAS really existed, the U.S. would be awash with women disabled with depression, filled with guilt, sexually dysfunctional, plagued with drug and alcohol problems. If PAS really existed, it would be entirely obvious. It doesn't exist. Belief in the existence of PAS is fantasy-based, brought to you by the same people who have a fantasy-based view of abortion and the Bible.

There is some evidence, however, that abortion causes a certain type of severe mental problem — mostly in men. For example, it's driven some religious leaders to the insane belief that the life of a zygote is more important than the life of a woman, that a zygote the size of a grain of salt is a genuine human being, etc.

In this chapter we've discussed some things you ought to know if you're considering abortion. If you're not considering abortion, there's only one thing you need to know: it's there if you ever need it. It's there if you're ever raped. It's there if a pregnancy someday threatens your life. It's there if someday you're pregnant but must give priority to the children you already have. It's there if you're pregnant but are not yet ready to shoulder the enormous responsibilities of being a parent.

If you're a woman, then abortion affects you even if you've never had one and never will. When abortion is legal and available, a woman gives birth because she wants to. When it's not available, a woman, all too often, gives birth because some men say she must. Your freedom to choose abortion, whether you use it or not, means you live in a better world. But that world is increasingly under attack, as we discuss in the next chapter.

Chapter 8

Abortion Battles

The popular CBS news program *60 Minutes* aired "God, Women and Medicine" on December 10, 2000. The program told the story of Kathleen Hutchins, whose water broke when she was fourteen weeks pregnant. Kathleen lived near Manchester, New Hampshire, where Elliot Hospital had recently merged with the Catholic Medical Center. Doctors said Kathleen's fetus had only a one in fifty chance of survival and recommended abortion. Without it, Kathleen risked infection, hysterectomy and possible death. Elliott Hospital policy had previously allowed medically necessary abortions, but now Elliott was associated with a Catholic hospital so policy had changed. Elliott administrators refused to authorize an abortion for Kathleen. From the show:

Morley Safer: "Elliot Hospital did allow abortions in cases of medical necessity. So Hutchins' doctor, Wayne Goldner, made an official request to the head of his department."

Dr. Wayne Goldner: "She said, 'I don't think I can get this through. I don't think we can perform this at the hospital without running into interference by the church and the board.'"

Morley Safer: "But the closest hospital he could send her to was 80 miles away and she had no means of getting there. So Dr. Goldner tried once again to schedule an abortion. Hospital officials said no."

Dr. Wayne Goldner: "They said, 'Just wait for her to get infected and bring her right in.' I said, 'Well, the whole purpose of this is not to get infected, to avoid the problem.'"

Morley Safer: "They were willing to take the risks."

Kathleen Hutchins: "They basically told me to sit at home and wait and see what happens."

Morley Safer: "They told you to go home and wait. Wait for what?"

Kathleen Hutchins: "Just wait for basically the fetus to die inside of me, and then something could be done about it."

Morley Safer: "That was something that both Ms. Hutchins and Dr. Goldner were unwilling to do. In the end he hired a car and driver to take her the 80 miles for an abortion. He says that religious doctrine clearly interfered with his patient getting the best medical care."

In 1973 the Supreme Court of the United States ruled that abortion was not the government's concern but was between a woman and her doctor. Thus began the battles over legal abortion that continue to this day. The same year, Congress passed the Church amendment, which included a "conscience clause" that allowed religion-controlled health institutions to refuse to perform abortions. Since then Congress and state legislatures have written more "conscience clauses" and "refusal clauses" into law. These clauses allow "faith-based" institutions to refuse not just abortion but other medical procedures that in the opinion of the institution conflict with their religious dogma.

The Church amendment was named after Senator Frank Church. But it might just as well be named for the churches and religions it was meant to please. It was designed so that religion-controlled medical institutions wouldn't have to perform medical procedures that violate their beliefs. But it applies to more than abortions, much more. For instance, in the U.S., Catholic hospitals must follow the seventy-two edicts of the ERDCHS (the "Ethical and Religious Directives for Catholic Healthcare Services"), written by the U.S. Conference of Catholic Bishops. In a Catholic-controlled hospital, the ERDCHS controls what medical care you can and can not receive. And the ERDCHS forbid much more than abortion.

Your grandmother is in a hospital, in a vegetative state. With her Living Will in hand, your grandfather tells her doctor: "She's said she'd never want to be kept alive in a condition like this. It's in her Living Will, too. I want you to remove her feeding tubes." (ERDCHS #24: *The institution ...will not honor an advance directive that is contrary to Catholic teaching.*) "Sorry," says the doctor, "but this is a Catholic Hospital and our Bishop would consider removing her feeding tubes as euthanasia. We can't do 'mercy killing' here." (ERDCHS #60: *Euthanasia is an action or omission that of itself or by intention causes death in order to alleviate suffering. Catholic health care institutions may never condone or participate in euthanasia or assisted suicide in any way.*)

Feeding tubes are "morally obligatory" for a patient in a persistent vegetative state, said Pope John Paul II. So your grandmother gets the feeding tubes, whether she wants them or not, whether her husband or family wants them or not, even if she specifically rejected them when she drew up her Living Will. In a Catholic hospital, a legal document like a Living Will is overruled by Church dogma.

A woman's cancer is treatable with chemotherapy. But chemotherapy will kill all her eggs and make her sterile. She and her husband want to have children some day so she asks her doctor to do an IVF procedure, to extract some of her eggs and freeze them after her husband fertilizes them. When she is well, she and her husband hope to re-implant some eggs and have a healthy child. The religion that runs the hospital doesn't approve of IVF, so her doctor refuses.

In Catholic and other religion-controlled hospitals, a woman who doesn't want any more children is denied abortion, tubal ligation, and even contraception. She may be refused prenatal testing if the hospital fears that negative test results would lead to abortion. On the other hand, a woman who wants to have children will be denied artificial insemination and IVF (in vitro fertilization). In a Catholic hospital, a woman doesn't control her own child-bearing. Even if she isn't Catholic.

You might suppose if a person doesn't like the policy of their neighborhood religion-controlled hospital they should just go somewhere else. Things are not so simple, for two reasons.

First, sometimes a religion-controlled hospital is the only one in the area. For example, there are over 500 Catholic hospitals in the U.S. compared to about 5,000 community hospitals. In ninety-one regions of the U.S., the Federal Government designates a Catholic hospital as the "sole provider" because no other hospital is close enough. Seventy-five percent of these hospitals are in regions where Catholics constitutes a small percentage (less that 25%) of the total population. If you live in one of these regions your tax dollars support a medical institution that denies you certain legal services based on their religious beliefs, not yours. In fact, one in five hospital beds in the U.S. are controlled by a religion, mostly Roman Catholic along with some Seventh-Day Adventists and Baptists. Religous groups control four of the ten largest U.S. health care systems. As one example, Ascension Health describes itself as "the nation's largest Catholic, nonprofit health system with more than 100,000 associates and health facilities in 20 states and Washington, DC." So, depending on where you live, it may be difficult or impossible to avoid a religion-controlled hospital. Kathleen Hutchins, who lived in New Hampshire, had to travel eighty miles.

The second reason is that even if your hospital is nonsectarian, it may be affiliated with a Catholic hospital in some way. If it is, then Vatican rules may prevail. You can avoid going to St. Teresa's Hospital but what if Main Street Hospital is now affiliated with St. Teresa's? Avoid going to Main Street, too? But what if they're the only two hospitals in town? And what if you go to Main Street not knowing it's affiliated with St. Teresa's?

A non-sectarian hospital and religion-controlled hospital can affiliate or merge with little or no publicity or community notification. Local community leaders, doctors, employees and patients may find out later, sometimes much later. The Bay Care consortium of Tampa Bay, Florida, for example, was created out of two Catholic and six nonsectarian hospitals. The local community didn't learn the impact of religious dogma on medical services until a year later. In New York a bill was proposed requiring the Department of Health to notify local communities of mergers between Catholic and non-sectarian hospitals. The bill was opposed by Catholic Bishops, and defeated.

You've been raped. You're sitting in a hospital Emergency Room. You tell your doctor you're terrified the rape has left you pregnant. You ask if there's anything she can do. She says: "I'm sorry but this is a Catholic hospital and we aren't allowed to provide emergency contraception." Emergency contraception is time-critical; it works best taken within three days of intercourse. It may work as many as five days after intercourse. So if access to emergency contraception is delayed long enough, access is effectively denied. Denying emergency contraception to a woman who is impregnated by rape leaves her two choices: have an abortion or bear her rapist's child.

The ERDCHS directive against emergency contraception is worth a detailed look. ERDCHS #36: "A female who has been raped should be able to defend herself against a potential conception from the sexual assault. *If, after appropriate testing, there is no evidence that conception has occurred already, she may be treated with medications that would prevent ovulation, sperm capacitation, or fertilization.* It is not permissible, however, to initiate or to recommend treatments that have as their purpose or direct effect the removal, destruction, or interference with the implantation of a fertilized ovum." This directive has a fantasy quality because the "appropriate testing" it mentions doesn't exist; no medical test determines if conception has occurred. As long as the egg is still in the fallopian tube no test can determine if it's fertilized or not. Available tests are useless until a fertilized egg implants in the uterus, which occurs six to ten days after intercourse. But six days after intercourse, emergency contraception is useless. In other words, the directive makes no sense medically. But, then, it was written by bishops not physicians.

You've been raped. You're sitting in a hospital Emergency Room. You tell your doctor you're terrified the rape has left you pregnant. You ask if there's anything she can do. She merely shakes her head. "No," she says, and nothing else. She doesn't say: "I'm sorry we don't provide emergency contraception at this hospital" because that tells you that emergency contraception exists and maybe you can get it somewhere else. And saying that can get her fired from a Catholic hospital because of a rule against "material cooperation," i.e., helping a person to get a forbidden procedure somewhere else. In a Catholic hospital, doctors, nurses and anyone else in the hospital can't provide certain services — and they can't even *mention* those services and refer you to another facility. If they do, they've violated the rule against material cooperation and risk losing their jobs because the ERDCHS is a set of directives, not suggestions. Any employees who fail to follow them may lose their job. ERDCHS #5: *Catholic health care services must adopt these Directives as policy, require adherence to them within the institution as a condition for medical privileges and employment.* You might expect that a Catholic hospital won't provide condoms and a Catholic-managed health care plan won't pay for them. But a physician who follows the ERDCHS cannot even mention condoms as a way of protecting against HIV and AIDS. Reportedly, a physician was fired from Catholic-affiliated New York Medical College merely for saying that legal abortion is "the law of the land."

Hospitals are not the only place where religion may control your health choices. In 1997, Congress allowed religion-managed health plans to refuse Medicaid payments that "provide, reimburse for, or provide coverage of a counseling or referral service if the organization objects to the provision of such service on moral or religious grounds." Taking advantage of the law, Fidelis Care, the sixth largest Medicaid managed health care program in New York City, refuses to pay for abortion, sterilization, birth control, and family planning counseling. In the U.S. all these services are legal. And Medicaid is a Federal program supported with tax dollars. Nonetheless, the 26,000 people covered by Fidelis Care are denied legal medical services that are forbidden by Catholic Bishops. If a woman wants those services, she must go "out of network" and pay for it herself. Many of Fidelis Care's 26,000 customers are poor women who can't afford to pay for it themselves. One of the things a poor woman may have to go "out of network" for is an operation to save her life — if the operation involves the death of a fetus. ERDCHS #45: "Abortion (that is, the directly intended termination of pregnancy before viability or the directly intended destruction of a viable fetus) *is never permitted.* ...Catholic health care institutions are not to provide abortion services."

So suppose you're sick but your life can be saved. But suppose saving your life requires ending your pregnancy, requires killing something the size of a raisin, a grape, or perhaps a lemon. Doing that is a sin. So instead you're allowed to die. Here's a excerpt from "Moral Problems in Hospital Practice," a book with a Catholic Imprimatur published in 1947.

> "Question: If it is morally certain that a pregnant mother and her unborn child will both die, if the pregnancy is allowed to take its course, but at the same time, the attending physician is morally certain that he can save the mother's life by removing the inviable fetus, is it lawful for him to do so?
> Answer: No, it is not. Such a removal of the fetus would be direct abortion."

The fetus is inviable, i.e., incapable of surviving, doomed, yet a "direct" abortion is not allowed, even to save your life. As one Catholic priest writes: "Two natural deaths are a lesser evil than one murder."

But if you're lucky you may not die, due to a loophole which Catholic theologians call the principle of "double effect." Here's an illustration of the principle of double effect. Case one (single effect): a man wants to kill people so he poisons a community's water supply. If ten people die then the man is guilty of ten murders. Case two (double effect): a man owns a chemical company and wants to increase profits. Legitimately disposing of chemical waste is costly, so instead he dumps poisonous chemical waste into the water supply — for the primary purpose of raising profits, but with the inescapable secondary effect of killing people. The man knows full well that dumping the waste will kill some people, but killing people is not his motive; saving money is his motive. Killing people is merely a secondary effect. If ten people die, is he guilty of ten murders, as in the first case? The principle of double effect says the man is not guilty of murder.

Normal courts of law don't use the principle of double effect, of course, but the Catholic Church uses it to allow some abortions that it calls "indirect" abortions, while still forbidding ordinary "direct" abortions. A direct abortion according to the Church is an operation whose primary purpose is the death of a fetus. For example, as long as Kathleen Hutchins wasn't infected and her life wasn't in danger, removing her fetus would be a direct abortion. But once she was infected, once her life was in danger, an operation to save her life that also happened to remove the fetus might be an indirect abortion.

As another example, suppose you have a tubal pregnancy. In a tubal pregnancy the fetus is doomed; such a pregnancy never ends in a healthy birth. And if the fetus isn't removed, it eventually bursts your fallopian tube; you hemorrhage and may die. But removing the fetus before it grows large enough to rupture your fallopian tube and cause you to hemorrhage might be a direct abortion. On the other hand, if you wait until the fetus causes your fallopian tube to burst, if you wait until you are hemorrhaging and your life is in danger, then the Church allows an indirect abortion. Specifically, ERDCHS #47 allows operations primarily aimed at saving a woman's life, even if the fetus dies as a consequence. And an operation that removes your hemorrhaging fallopian tube is primarily for saving your life, so it's permissible. The fetus happens to be in your fallopian tube (and, in fact, is what has caused the hemorrhaging) but the operation isn't primarily to kill the fetus so it complies with the ERDCHS. In the eyes of the Catholic Church the operation is a permissible indirect abortion. But an operation aimed at removing the doomed fetus before your life is threatened is a direct abortion.

The principle of double effect can cost you a fallopian tube. Here's how. Imagine you have a tubal pregnancy. If you go to a non-sectarian hospital, there are two treatments. Early in your pregnancy you can be given Methotrexate, which causes a spontaneous abortion. Later in your pregnancy, there's an operation that removes the doomed fetus but leaves your fallopian tube intact. With either method, you will still be able to get pregnant in the future. But both of these methods have as their object the removal of the doomed fetus, and to the persnickety male moralists of the Catholic Church that's a no-no. So if you happen to be in a Catholic hospital, the physician may remove all or part your fallopian tube along with the fetus, not to satisfy any medical requirement but to satisfy a requirement of religion; specifically, removing only the fetus is in the eyes of the Church an impermissible direct abortion, but removing the fetus along with some or all of your fallopian tube is an allowable indirect abortion. The operation you receive may be dictated not only by medical necessity but also by theological conformity. It doesn't matter if you're Catholic or not. You must play by the rules dictated by the Vatican. You are denied the medicine that you and your doctor think best if the Vatican doesn't allow it. You are denied the operation you think best unless the Vatican agrees. Religious dogma trumps medical necessity.

When a hospital is a private institution of some religion, run entirely on Church funds, refusing certain medical services might seem acceptable. But in the U.S. many religion-controlled hospitals are funded with public money. Often, half or more of the funding of such a hospital comes from the taxpayers. A strictly private hospital can refuse services that violate its religious principles. But should hospitals that are supported with your tax dollars have the same right?

Religion-controlled hospitals receive about forty-five billion a year from the Federal government in the form of Medicaid and Medicare payments. Government grants provide millions more. And private health plan companies add billions to that. Where does money for Medicaid, Medicare and government grants come from? From your tax dollars. From you. And where do health plans get their money? From you and your employer. And who gets to control how those dollars are spent? Who gets to control what medical care you can and can not receive? The Vatican, if you happen to be in a Catholic-controlled hospital or enrolled in a Catholic-controlled health plan.

Recently the battle has spread from hospitals and health managed care plans to your local pharmacy. Women have walked into a pharmacy and been refused a medicine prescribed by their doctor, because the medicine violated the religious beliefs of the pharmacist. In New York a woman with a serious medical problem was advised not to get pregnant for a few years. Her doctor gave her a prescription for birth-control pills; her pharmacist refused to fill it. In Texas, a woman was refused birth control pills because the pharmacist "didn't personally believe in birth control. It's not right." Worse, laws have been passed in some states (Mississippi, South Dakota and Arkansas, to name three) that protect pharmacists who refuse to fill birth control prescriptions for religious or moral reasons.

Your pregnancy is going badly. The doctors believe your life is in danger. Medically, they should end your pregnancy. But you're at a Catholic hospital and Catholic hospitals don't do abortions. Kathleen Hutchins had to travel eighty miles for an abortion. But what if all available hospitals had been too far? What if all hospitals within a few hundred miles were under the ERDCHS? In 2004 Congress and state legislatures were considering bills that allow health providers to refuse any service against their religious principles *without regard for the patient's health or life.* As I write this, none of the bills have passed. I hope the same is true as you read this.

And it's going to get worse before it gets better. Because in recent years, anti-abortion groups have taken control of the U.S. government. The list of their actions against abortion, contraception, family planning and sex education is a long one. They have removed family planning and contraception coverage from various Federal health programs; appointed religious hysterics to government committees; made it a federal crime to transport a minor across state lines for an abortion without parental consent; reinstituted a global gag rule on international family planning assistance; removed medically-accurate condom information from some Federal web sites; withheld funds for the United Nations Population Fund; and more.

They have also increased funding for abstinence-only sex education. What is abstinence-only sex education? It is sex "education" that exaggerates or lies about the health risks of sexual activity, contraception and abortion; pretends that masturbation is abnormal; avoids the subject of contraception or pretends contraception doesn't work; has little or nothing to say about safe sex practices, about how to avoid unintended pregnancy or sexually transmitted disease; pretends abortion leads to various medical problems; has no information about how to actually obtain contraceptives.

In fact, many abstinence-only programs have hardly any medically-accurate information at all; they are riddled with misleading or incorrect information, as was described in at least one congressional report. Abstinence-only sex "education" is, in fact, fantasy-based education, promoted by the same people who have fantasy-based beliefs about PAS and the Bible's view of abortion.

Over last two decades more and more U.S. school educational programs have become abstinence-only. Because of such programs, teens in the U.S. have less access than teens in Europe to sane sex information and contraception. And the U.S. has the teen pregnancy rate to show it. Studies show that students who have abstinence-only education aren't less likely to have pre-marital sex, but they are less likely to use contraception when they do. In other words, they are more likely to get pregnant. So among the world's developed countries, the U.S. has the highest rate of abstinence-only sex education and the highest rate of teen pregnancy. Teens in the U.S. are *twice* as likely to get pregnant than teens in England and France. In the U.S., only California refuses federal abstinence-only funds. California's reality-based sex education programs have helped reduce its rate of teen pregnancy to close to some European countries.

If you're a woman of child-bearing age, the abortion battles concern you. If you're a woman who has not yet reached child-bearing age, the abortion battles concern you. If you're past your child-bearing years but care about your daughters and granddaughters, the abortion battles concern you. If you care about your wife, your daughter, your mother, your sister, the abortion battles concern you.

If you don't think that a family should ever be forced to see a loved one kept alive in a vegetative state, the abortion battles concern you.

Chapter 9

The Larger War

The abortion battles are part of a war that has raged for centuries. There have been many battles in this war — battles over disease and insanity, over "witches," over childbirth pain, over contraception, over masturbation and, today, over abortion. But the battles all have an element in common that shows they are all part of a single, larger war.

What links all the battles together? The usual answer is that each battle is over the same basic question: is religion a genuine source of earthly knowledge? In particular, is it a genuine source of medical and scientific knowledge?

Medicine treats the body; science, the material world; religion, the spirit. It would be disastrous if scientists and physicians meddled with religious dogma. The meddling of religious leaders with medicine and science has been no less disastrous. Books can (and have) been written about the subject, but here's a brief summary. We'll begin about 1,700 years ago, when the battle was being fought over disease and insanity.

What causes disease and insanity? Some ancient physicians believed in natural causes and were progressing towards a scientific method of treatment. Religion had another cause: the supernatural. Augustine, for example, taught that "All diseases of Christians are to be ascribed to demons." There are solid scriptural justifications for such a view of disease. In Matthew 9:32, a devil causes dumbness. In Matthew 12:22, a devil causes dumbness and blindness. A devil causes insanity in Matthew 17:18; numerous devils cause it in Mark 5:3:13. And how should we cure disease and insanity? Well, supernatural problems obviously require supernatural cures. So some "cures" might be relics (the bone or possession of a dead saint), holy water, and miraculous pools and streams. There is a solid scriptural basis for these cures, too. In Acts 19:11-12, a handkerchief and apron touched by the Apostle Paul cure disease and drive out devils. In II Kings 13:21, the bones of the prophet Elisha bring a dead man back to life. In II Kings 5:13, washing seven times in the Jordan River cures a man. In John 9:7, washing in the pool of Siloam cures a man of blindness.

The battle over disease and insanity in the ancient world was fought between those who believed disease and insanity had natural causes, and those who believed the causes were supernatural. When the battle was over, the supernatural side had won. Western civilization turned from a natural to a supernatural view of disease and insanity. Over the next fifteen centuries, the old naturalistic view was suppressed; much genuine knowledge and progress concerning disease and insanity were lost. As a result how many people suffered needless pain and death? The number is incalculable. And enormous.

A thousand years after Augustine, the battle was over witches. In the 1300's religious people decided to murder witches, a practice that also has a solid scriptural basis: Exodus 22:18, *Thou shalt not suffer a witch to live.* They also decided cats were instruments of witchcraft, evil and the devil. So they decided to kill all the cats they could find. If you lived in this world and liked cats or, worse, actually had one or two pet cats, you risked being tortured and then killed. In that world, being a woman was hazardous, especially an old woman, especially an old haggard woman, especially if she owned a cat or two, especially if personality or mental illness caused other people to regard her as strange. People in that world were killing "witches" and cats — and any people who defended them. At first, religious hysterics won that battle: they were well on their way to eradicating "witches" and cats from all of Europe. But as the cat population dwindled, the rat population increased. More rats meant more carriers of the bubonic plague, also known as the "Black Death." After the Plague had killed an astounding one-quarter of the entire population of Europe, the religious hysterics decided maybe killing cats wasn't such a good idea.

In 1591 the battle was over the pain of childbirth. Suppose you lived in Scotland then. And suppose you were giving a painful birth and begged for something to relieve the pain. Then your life was in danger but not necessarily from childbirth. In 1591 Euphanie Macalyane begged her midwife for something to easy the terrible pain of childbirth, and she gratefully took what the midwife offered. James VI, then King of Scotland, later King of England, was enraged that any woman would attempt to evade God's punishment for childbirth as recorded in the Bible: "I will intensify the pangs of your child-bearing; in pain shall you bring forth children." So the religious King James had Euphanie Macalyane brought to Castle Hill of Edinburgh and burned alive. (Later, when he became King of England, he authorized the English Bible translation known as the King James Bible.) The battle against childbirth pain was lost.

We've already discussed the famous battle over whether the earth revolves around the sun. People who believed in a fixed, unmoving earth initially won the battle. For example, Galileo, considered by many the "Father of Modern Science," was brought to trial by the Church and forced to recant his beliefs. And in 1633 his famous book "Dialogue Concerning the Two Chief Systems of the World" was placed on the Catholic Church's Index of Forbidden Books. But the truth eventually won. In 1824 Galileo's book was removed from the Index. In 1966, the Index itself was abolished. In 2000, Pope John Paul II officially apologized for the trial of Galileo.

In 1752 the battle was, of all things, over the lightning rod. In a famous experiment, Benjamin Franklin used a kite in a thunderstorm to show that lightning is a form of electricity. Franklin then invented the lightning rod, which protects buildings against lightning. But religion for centuries had taught that lightning was God's punishment. So it called Franklin's invention "the heretical rod" and opposed its use. Some clergymen said the 1755 Boston earthquake was God's punishment for using the lightning rod. In those days, churches were commonly built on a hill overlooking the town. The church's height and long spires, reaching to heaven, made it a perfect target for lightning. Why was God's own home, people often wondered, struck so often by lightning? Their wonder increased as homes protected by "the heretical rod" were spared while God's own home continued to be smitten. Slowly, churches began adding lightning rods. Slowly, the battle was won in favor of the lightning rod.

In 1795, battle was joined again when Dr. Edward Jenner of Gloucestershire, England, discovered smallpox vaccination. There was religious opposition. Vaccination was denounced as "defiance to Heaven itself, even to the will of God." Ninety years later, in 1885, a smallpox epidemic struck Montreal, Canada. The Catholic Church still saw disease as a punishment from God, and one priest preached: "if we are afflicted with smallpox, it is because we had a carnival last year, feasting with the flesh, which has offended the Lord; ...it is to punish our pride that God has sent us smallpox." Catholic Bishops opposed vaccination and advised increased prayers, especially the rosary. A special procession in honor of the Blessed Virgin was organized. Many Catholics refused to be vaccinated. When Catholics died in high numbers, surviving Catholics decided that vaccination wasn't such a bad idea after all. Eventually the battle was won in favor of vaccination.

In 1847 the battle over childbirth pain began again, again in Scotland, again in Edinburgh. Dr. James Simpson, professor of medicine and midwifery, discovered that chloroform was an effective painkiller for a woman in childbirth. By 1847 religion had largely lost its power to torture and murder, so all it could do was fume and fuss: "What a Satanic invention! What a shame upon Edinburgh!" said the Scottish Calvinist Church.

Pamphlets were sent to doctors condemning the discovery: "To all seeming, Satan wishes to help suffering women but the upshot will be the collapse of society, for the fear of the Lord which depends upon the petitions of the afflicted will be destroyed." About six years later, Queen Victoria demanded a painkiller while giving birth; that ended the battle in England in favor of painkillers for childbirth.

The battle over contraception has raged for centuries and is still in progress today. Religion-administered Medicaid plans deny contraception coverage. Pharmacists refuse to fill prescriptions. Yet the Bible mentions contraception once, at most. In Genesis, Onan's brother dies; custom dictates that Onan must impregnate his brother's widow. But Onan apparently has some second thoughts about following custom for he withdraws before the act is completed (i.e., practices coitus interruptus). So God kills him. The Bible doesn't say if God's anger is towards Onan disobedience to duty or the coitus interruptus itself. The traditional Jewish teaching is that the sin of Onan was disobedience. But early Christian teachers decided it was contraception. They claimed the Bible condemns contraception, a claim that makes as little sense as the claim that the Bible condemns abortion. Later, when the Church wanted to condemn masturbation they used the story of Onan for that, too; the word "Onanism" thus became a synonym for masturbation. Today, religious people who oppose abortion base their opposition on misinterpreted Biblical verses. In the past, the same type of people based opposition to contraception and masturbation on misinterpreted Biblical verses.

Christianity's opposition to contraception dates, at least, to Augustine, who condemned it and taught that sex was unlawful and evil when it could not lead to children. Throughout the centuries other Christian leaders followed Augustine. Martin Luther said contraception was as evil as sodomy. John Wesley said it was "unnatural and destroys the souls of those who practice it." John Calvin said contraception was "the murder of a future person." (Calvin's horror over the "murder of a future person" didn't stop him from burning to death an actual, living, breathing person by the name of Michael Servetus, due to a religious disagreement.) To today's religious hysteric, abortion is murder. To religious hysterics of the past, using a condom was murder (of a future person). But religious hysterics have often accepted with aplomb the killing of actual human beings, in war or through capital punishment. During the Inquisition, they accepted the torture and murder of thousands of human beings.

In 1873 the battle over contraception was still being fought. In the U.S., the Comstock Act banned contraception from interstate commerce. In 1877, a woman and man were tried for selling a pamphlet about contraception. The first U.S. birth control clinic opened in 1916 — and was closed by the police shortly later. The same year, the Vatican declared that a wife must resist her husband "as she would a rapist" if he uses a condom. But the tide began to turn in 1930 when the Episcopalian Church granted acceptance to contraception; many other Protestant denominations eventually followed suit. By 1960 only five U.S. states still had laws against selling contraceptives. But in some New England states, physicians were not allowed to distribute contraception information to patients, even if the patients requested it. In 1965 the U.S. Supreme Court decided banning contraception was illegal. That ended the legal battles against contraception in the U.S. But the Catholic Church still refuses to accept contraception. In 1993, Pope John Paul II said contraception was "intrinsically evil." In fact, in 1988 he said that a man with HIV/AIDS should not use a condom to have sex with his wife. (A man with HIV/AIDS who refuses to use a condom with his own wife is a classic example of the religious hysteric — someone whose innate good sense has been warped by religion.) In some parts of Africa HIV/AIDS infects as much as ten percent of the population. Yet in 2000 the Archbishop of Uganda condemned the use of condoms in fighting HIV/AIDS. In 2005 South African Cardinal Wilfred Napier said there is "no evidence" that condoms help prevent HIV/AIDS. He also said that telling young people about condoms encourages them to engage in sexual activity. (A hyper-moral attitude towards sexuality is another trait of the religious hysteric.)

Apparently the battle over contraception is still going on.

In the 1700 and 1800's the battle was over masturbation. About 1712 a Dutch theologian published *"Onania, or the Heinous Sin of Self-Pollution, And All Its Frightful Consequences, In Both Sexes, Considered: With Spiritual and Physical Advice To Those Who Have Already Injured Themselves By This Abominable Practice."* Masturbation, said the misinformed theologian, causes digestive problem, vomiting, nausea, weakness, paralysis, and ten or fifteen other serious ailments.

In 1879 Dr. J.H. Kellogg, M.D., wrote "Plain Facts for Old and Young." Dr. Kellogg was no obscure crank. Rather he sat for twelve years on the Michigan State Board of Health, was a member of various professional societies, and upon his death received tributes from such notables as John D. Rockefeller, Jr., and President Herbert Hoover. Kellogg's book "Plain Facts" is anything but. In it the doctor gives some signs of masturbation for the concerned parent: "confusion of ideas; round shoulders; weak backs; pain in the limbs, and stiffness of the joints; paralysis; lack of development of the breasts in females after puberty." He also writes, "Eating clay, slate-pencils, plaster, chalk, and other indigestible articles is a practice to which girls who abuse themselves are especially addicted." Masturbation, says Dr. Kellogg, results in "impotency in the male, sterility in the female," not to mention "heart disease, diseases of the nervous system, epilepsy, cancer, idiocy, suicide, insanity and piles." So he advises parents to be vigilant. But what's a concerned parent to do? The good doctor suggests, "bandaging the parts," "covering the organs with a cage" or "tying the hands." (In the 1800's there was a lively commerce in masturbation control devices such as "penis cases, sleeping mitts, bed cradles to keep the sheets off the genitals, and hobbles to keep girls from spreading their legs.") If all that fails, Dr. Kellogg observes that circumcision is "a remedy that is almost always successful in small boys... The operation should be performed by a surgeon without administering an anesthetic, as the brief pain attending the operation will have a salutary effect upon the mind, especially if it be connected with the idea of punishment."

Medical misinformation about masturbation persisted into the 1900's. For example, in the early 1900's Lord Baden-Powell, who founded the Boy Scouts, said the boy who masturbates "becomes feeble in body and mind and often ends in a lunatic asylum."

Earlier we asked what all the battles in the larger war have in common. We said that the usual answer is that the battles were over one basic question: Is religion a genuine source of earthly knowledge? But this answer doesn't explain the enormous battles over contraception, masturbation or abortion, none of which the Bible condemns, forbids or even mentions. And it doesn't fully explain the horrible witch hunts of the Middle Ages. (True, the Bible does say, *Thou shalt not suffer a witch to live.*" But the quote occurs in the Old Testament. Christians ignore many other Old Testament laws. Why did they decide to follow the law against witches?) And it doesn't explain the sadism. Even if Euphanie Macalyane had disobeyed some law of God, the New Testament says to forgive and not to judge. So the perverse sadism she suffered at the hands of King James can't be blamed on the Bible. So what, in fact, is the common thread that runs thought all the battles?

Chapter 10

Speculation

You've now reached the last chapter of the book. Congratulations. And thanks for taking the time to read it. I hope you feel it's been time well-spent. Though I've offered numerous opinions, I view the previous chapters as primarily fact-based. The reasons for abortion, the overview of fetal development, the history of abortion and religion: these and many of our other topics have a foundation in fact. Though I've omitted footnotes to make an easier read, footnotes could be given. This chapter will be different. The views offered here are mostly opinions, mostly my own opinions. For instance, the explanation we'll see of the religious hysteric is merely what I believe to be true; I have no scholarly footnotes to support it. And the statements about the future are also opinion; like all guesses about the future, they are speculation. But speculation can be useful. For instance, our portrait of the religious hysteric will unify various facts from previous chapters; for once the mind of the religious hysteric is understood, the connections between his various actions and attitudes become clearer.

We ended the last chapter with a question: What is the common thread that runs thought all the battles? In my opinion, each battle of the larger war has been between sane, rational people and religious hysterics. We've mentioned the religious hysteric a few times but will now discuss him in more detail.

The hysteric is someone who is temporarily out of his or her right mind, someone who is in a state of temporary hysteria. As an example, consider Dr. Kellogg, who we met in the last chapter. The good doctor had enough intelligence to become a prominent physician. He sat on a state Board of Health and was a member of professional societies. No doubt, he was sane and in his right mind most of the time. But the topic of masturbation drove him to hysteria, a state where emotion overrules logic, where fear distorts facts, where without any evidence he "knew" the long-term effects of masturbation.

The person driven to hysteria by religious belief is a religious hysteric. Thus, if religion was behind Kellogg's mania about masturbation then he was a religious hysteric. As another example, consider the Catholic bishops who repeatedly transferred child-molesting priests from one parish to another. Those bishops were not fools, not natural born idiots. Rather they were intelligent men making decisions while "operating under the influence of religion." Anyone with a modicum of good sense would have kept predator priests away from defenseless children. But the bishops repeatedly shuffled them from one parish to another, where the priests sexually abused one child after another. What caused the bishops' colossal lack of good judgment? Religious hysteria. The issue of child-molesting priests temporarily caused them to become religious hysterics, to temporarily take leave of their senses.

One trait of the religious hysteric is that he "knows" without any supporting evidence. Thus, the Catholic hierarchy "knew" the right thing to do was to transfer the child-molesting priest. Thus, the anti-abortion religious hysteric "knows" that abortion leads to breast cancer, and "knows" that PAS (Post Abortion Syndrome) exists. Thus Dr. Kellogg "knew" that girls who masturbate eat "clay, slate-pencils, plaster, chalk, and other indigestible articles." Such "knowledge" is obviously not fact-based but fantasy-based. Were the religious hysteric in his right mind he would look for facts before making decisions. But he isn't in his right mind, so facts aren't necessary; he already knows with absolute certainty. So he can ignore facts entirely. More commonly, he finds a few facts that support his beliefs and ignores any others.

Because few or no facts support his opinions, the religious hysteric can't argue his beliefs rationally. So he uses emotion. For example, to argue against abortion he calls the microscopic zygote a "baby" and even a "child." (Why stop there? Why not a "teenager" or "senior citizen"?) Historically he labeled birth control as murder, too. But though he is hysterical in his opposition to the "murder" of raisin-sized embryos, or the "murder" caused by birth control, the religious hysteric often easily accepts the murder of actual human beings (the "witches" of the Middle Ages, a hundred thousand civilians in war). He'll piously talk about the sanctity of human life and, in the same sentence, calmly describe the necessity of bombing an abortion clinic. Clearly, he is not in his right mind.

The beliefs of the religious hysteric aren't based on fact and often aren't based on Scripture either. He justifies his most fervent beliefs (for example, that the Bible forbids masturbation, birth control and abortion) with vague or misinterpreted Scriptural verses. Though he may, in fact, be wrong, he is always certain because in his own distorted mind God or Scripture tells him so. People in their right mind realize their own opinions may be wrong. But when it comes to his pet phobias, the religious hysteric has no fact-based opinions. Rather, he has fantasy-based delusions that he mistakes for absolute truth. The religious hysteric is certain his beliefs come from God, so he sees contrary beliefs as coming from the Devil. And he sees people with contrary beliefs as not merely having a different point of view, but as evil, as in league with the Devil. And if evil resists, force must be used. Thus the Crusades and Inquisition. Thus, religion's burning of "heretics" (those with contrary religious beliefs) throughout the ages. Thus, repressive laws that force everyone else to follow the religious hysteric's beliefs. And because he pictures himself in a fight with evil and the Devil, he has no hesitation using any methods, even dishonest or sadistic methods. Thus, the phony "abortion counseling centers." Thus, the medieval torture and murder of "witches." Thus, the wholesale murder of cats.

Of course, the religious hysteric doesn't see himself as dishonest, intolerant or sadistic; rather in his own deluded mind he's a humble servant doing God's work. No doubt the religious King James saw himself as merely doing God's work when he sadistically burnt Euphanie Macalyane to death. And because he sees himself as merely doing God's own work, the religious hysteric often has a calm acceptance of monstrous acts: using circumcision without anesthetic as a punishment for masturbation, the torture and burning of heretics and "witches," the bombing of abortion clinics. In his own eyes he is a highly moral person, yet another indication his religion has robbed him of his innate good sense.

Because the religious hysteric is often sadistic, he often pictures his God as sadistic, too. Thus, Hell, his God's eternal place of torture, where the religious hysteric sends most of the people who have ever lived: people who have never accepted Jesus as their personal savior (some Protestant denominations), people who died with unforgiven mortal sins like masturbation, contraception, abortion, missing Mass, and, some years ago, eating a hamburger on Friday (the Catholic Church).

The phobias of the religious hysteric usually have something or other to do with sex. The sexuality of the religious hysteric is a mixture of fearful repulsion and leering attraction; in other words, he has the sexual personality of a leering adolescent. For example, in the U.S. a hundred years ago, legislators wrote laws that outlawed non-vaginal sex. But they had such a juvenile embarrassment of sexuality that they just couldn't bring themselves to write the words "fellatio" and "cunnilingus." So they wrote laws that outlawed vague "crimes against nature." To the immature mind, words have power.

The religious hysteric feels threatened by sexuality. That's why he wants to have as little to do with sex as possible; that's why he says that the only valid use of sex is reproduction. He imagines having such an opinion of sex elevates him closer to the angels. In fact, it brings him down to the level of the lower animals. For it is the lower animals that use sex exclusively for reproduction. Among the lower animals the female is receptive to sex only when she is in heat and can get pregnant. But women don't go into heat like cats and dogs. God so designed the human female that she has sexual urges even when she's not ovulating. According to the religious hysteric, a woman should deny her sexual urges unless she can fulfill the "true" function of sexuality: reproduction. But, unlike the religious hysteric, most people realize that sex has a higher purpose than mere reproduction. Sex is primarily for the expression of intimacy and love, for affection, and even for fun. Its reproductive purpose is secondary.

It's obvious that the religious hysteric isn't truly religious. Rather he misuses religion to support and reinforce his hysteria. That's why he so easily twists Scripture to support his sexuality phobias (masturbation, contraception, abortion). He doesn't want to follow Scripture. Rather he wants Scripture to follow him, to support his beliefs. He doesn't want to be tolerant or forgiving towards people who don't share his phobias. He wants to feel such people are evil. So he uses Scripture to support his manias. But when Scripture says not to judge and to forgive, he easily ignores it.

In the U.S. studies show that anti-abortion people are often politically right-wing; they often oppose government social services, like pre-natal care for poor, pregnant women, or free school lunches for impoverished children. Such people are very concerned with the lives of pea-sized embryos; about the lives of actual children they aren't so concerned. They sancti-moniously oppose family planning services in countries where overpopulation is killing infants, children, and women. They oppose condom use in countries racked with AIDS. Though they'd deny it, their approach to population control is essen-tially Hitler's approach: no contraception, no abortion, let the fittest survive, let the rest die. In their own deluded minds they are close to God. In fact, they are well-fed, and even over-fed, people opposing population control in countries where chil-dren are starving. Can such people actually be close to God?

What causes someone to become a religious hysteric? The religious hysteric can be viewed as someone who, for one rea-son or another, fails to make a successful transition from the emotional life of the child to the emotional life of the adult. Emotionally, the religious hysteric is still a child.

A child believes with absolute certainty. The young child never suspects that Santa Claus doesn't exist. Santa Claus does exist; his parents told him so. So he *knows* it; he feels it in his heart. The normal adult eventually loses the certainty of the child, along with the need for that certainty. He is able to lead a full, satisfying life amid the uncertainty and doubt inherent in being human. But the religious hysteric never comes to terms with uncertainty and doubt. He remains in need of the abso-lute certainty of the child; he craves it. So he uses Scripture to give himself divine sanction for whatever he wishes to believe. Sometimes a particular belief happens to be in Scripture; some-times it is not. But since the religious hysteric misuses Scripture to give authority to what he believes, it doesn't matter if his pet beliefs are in Scripture or not. With the simplistic reasoning of a child, he twists Scriptural verses to support whatever he believes; verses that he doesn't wish to follow, he simply ignores. Then he solemnly decides that Scripture says what he wants it to say, that he is truly following Scripture, and is genuinely shocked and incredulous if anyone suggests otherwise. The religious hysteric is absolutely certain that his beliefs are right and approved by God; so he can find no explanation for people who disagree except they are "evil." Thus, the world for him is a simple-minded fight between "good" (whatever he wants to believe is true) and "evil" (contrary beliefs).

The religious hysteric craves the absolute certainty of childhood and the simple-minded good/bad worldview of childhood. So it's not surprising that the religious hysteric craves the sex-free world of childhood, too. Sexuality is powerful and can be scary, especially at first. So the young child looks for protection and assurance. I once read about a woman who had three children. Her oldest daughter was just coming to the realization that her mother had "done it." "How could you?" asked her oldest. "And three times!" "Well, honey," replied her mother with great wisdom, "if you want children bad enough you'll do anything." The mother's answer reassured the young girl; it protected her from a truth she was not ready to accept. The girl understood that without sex the human race would eventually die out, so she was willing to allow sex into her otherwise sex-free world, but only for reproduction. Most people eventually allow sex into their world for more than reproduction; they eventually come to terms with sexuality, one way or the other. When done in a sane, balanced way, sexuality is integrated as an enjoyable part of life. When done in a less that sane manner, sexuality becomes a dominating force, an obsession.

People who obsessively act on their sexual impulses have numerous indiscriminate relationships characterized by shallowness and deception. In the extreme case, they become rapists. People who obsessively deny their sexual impulses become prudish and puritanical, with a hypersensitive, hypermoral attitude towards all things sexual. And they often have an obsessive interest in the sexual activities of their neighbors. In the extreme case, they become religious hysterics and preach that sex is just for reproduction.

The religious hysteric's obsession with sex is apparent in various ways. To name one, the sins he worries about are all sexual sins. For example, an "immoral woman" is a woman who (in the opinion of the religious hysteric) indulges too easily in sex. In other words, for the religious hysteric "morals" and "sexual morals" is the same thing. For the religious hysteric, kindness, generosity, concern for others, truthfulness, and a thousand other virtues don't define morals; sexual behavior alone defines morals.

The religious hysteric never really comes to terms with his own sexuality, never successfully integrates it into his life. He remains extremely, sometimes obsessively, concerned with sex — not with indulging in sex, but with fighting his own sexual urges. And fighting everyone's sexual urges, too. Thus, the long history of oppressive, puritanical laws against premarital sex, contraception, and most other things that concern sex. And he often sees sex (and members of the opposite sex) negatively. Augustine is an interesting example of someone who never successfully integrated sexuality into his life. In his early days, he obsessively acted on his sexual impulses. Later in life, he became a religious hysteric and obsessively rejected them. Augustine expressed the attitude of religious hysterics throughout the centuries when he wrote: "Nothing is so much to be shunned as sex relations." and "Women should not be enlightened or educated in any way. They should, in fact, be segregated as they are the cause of hideous and involuntary erections in holy men." In fact, what is hideous is Augustine's view of half the human race.

Religious hysterics like Augustine greatly influenced Christianity at the time when it was still young, still formalizing its dogma. During the Middle Ages, religious hysterics took control of the Catholic Church and instituted the two-hundred year orgy of torture and murder called the Inquisition. All of the Inquisition's officials were men; most of its victims were women. Today religious hysterics are less powerful, but still deeply affect the Catholic Church's teaching about sex.

The war against religious hysterics has been in progress for thousands of years and will probably continue for thousands more. At present, religious hysterics are in power and are chipping away at your rights to abortion and contraception. Since we are speculating, it's tempting to ask: What does the future hold?

The future of abortion and contraception isn't clear. Perhaps someday you'll lose your right to abortion and even to contraception. Perhaps not. But one thing is clear: while particular forms of population control (i.e., abortion and contraception) are at risk, the future of population control itself is not.

The earth's population will be controlled, one way or the other. There will be population control, the only question is: what type? Post-birth or pre-birth?

Post-birth methods of population control are starvation, disease, and war. When a country can no longer support its growing population, some people starve or, weakened by malnutrition, die of disease. Or the country goes to war in an effort to provide for its people. We've seen how Germany went to war when it had (or, at least, believed it had) more people than it could support. We can only speculate what wars have been avoided and how many deaths have been avoided by China and India's intelligent use of pre-birth population control.

Pre-birth methods are contraception and abortion. Contraception doesn't kill human beings and neither does abortion if performed early enough. Yet people who absurdly call themselves "pro-life" favor post-birth over pre-birth methods of population control. Such people would deny that they favor any type of population control at all; but that, like so many of their other claims, is fantasy-based. Population control can not be avoided. In fact, both methods of population control are in use now, at this very moment. The pre-birth methods of contraception and abortion are in use throughout the world. But so are post-birth methods. The post-birth method of malnutrition-related illness kills about ten million people a year. The post-birth method of polluted drinking water kills millions of children a year and sickens many others, making them more susceptible to disease.

But even with these methods, the earth's population is increasing at a dangerous rate. While our population is mushrooming we are losing some of the land that grows food. Land overuse and misuse has destroyed the food growing capacity of a third of the earth's soil. At the present rate of increase the earth will have three billion more people by 2050. Yet some scientists doubt that the earth in the long term can support its present population of six billion people. Clearly, to avoid the horrors of starvation, disease and war, the earth is desperately in need of more pre-birth population control. But religious hysterics are doing everything they can to give us less pre-birth population control; they are doing everything they can to deprive you of your access to contraception, and the option of abortion should you ever need it.

Are you going to let them?

For more information
visit
http://WisdomOfAbortion.com